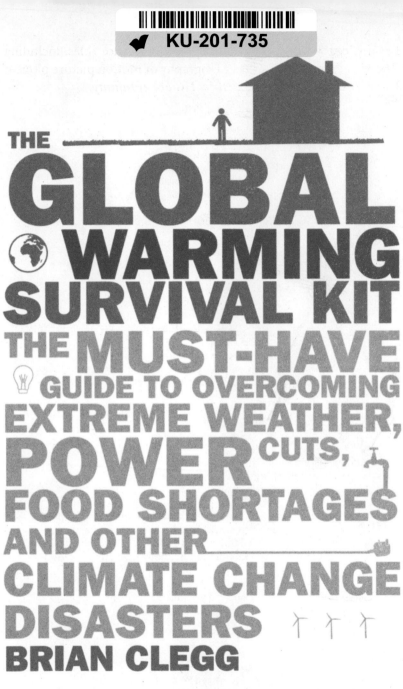

THE GLOBAL WARMING SURVIVAL KIT
THE MUST-HAVE GUIDE TO OVERCOMING EXTREME WEATHER, POWER CUTS, FOOD SHORTAGES AND OTHER CLIMATE CHANGE DISASTERS
BRIAN CLEGG

Doubleday

LONDON • TORONTO • SYDNEY • AUCKLAND • JOHANNESBURG

TRANSWORLD PUBLISHERS
61–63 Uxbridge Road, London W5 5SA
A Random House Group Company
www.rbooks.co.uk

First published in Great Britain
in 2007 by Doubleday
an imprint of Transworld Publishers

A CIP catalogue record for this book
is available from the British Library.

ISBN 9780385612609

Addresses for Random House Group Ltd companies outside the UK
can be found at: www.randomhouse.co.uk
The Random House Group Ltd Reg. No. 954009

The Random House Group Ltd makes every effort to ensure that the papers used
in its books are made from trees that have been legally sourced from well-
managed and credibly certified forests. Our paper procurement policy can be
found at: www.randomhouse.co.uk/paper.htm

Typeset in 11/13pt Minion
by Falcon Oast Graphic Art Ltd

Printed and bound in Germany by GGP Media GmbH, Pössneck.

2 4 6 8 10 9 7 5 3 1

For Gillian, Chelsea and Rebecca

ACKNOWLEDGEMENTS

My thanks to my agent Peter Cox and editor Susanna Wadeson for making the experience of writing this book so enjoyable.

Sea-level-rise maps courtesy of Weiss and Overpeck, the University of Arizona, reproduced with permission.

CONTENTS

ONE

TEETERING ON
THE BRINK

THE EARTH'S CLIMATE is changing. This is not news. The US National Academy of Sciences made their first study of global warming back in 1978. Although widespread acceptance that there is a serious problem took time to develop, the impact of climate change has now been studied for a good number of years – and the vast majority of scientists accept that this change is strongly influenced by human activity.

The UN added its support in the 2007 report from the Intergovernmental Panel on Climate Change (IPCC), stating that global warming is unequivocal fact, and that most of the rise since 1950 is most likely (with a better than 90 per cent confidence) to have been caused by human intervention. 'February 2 [2007] will be remembered as the date when uncertainty was removed as to whether humans had anything to do with climate change on this planet. The evidence is on the table,' said Achim Steiner, executive director of the UN Environment Programme.

Even the few who don't accept a man-made component admit that we are undergoing global warming. According to the IPCC, the world can look forward to centuries of climbing temperatures, rising seas and disrupted weather. All the evidence is that the world is warmer now than at any time in the past two millennia; if current trends continue, by the end of the century it will be the hottest it has been in two million years.

The ten warmest years on record have all occurred since 1990, and most of those were in the last decade.

There is a lot of talk about action to prevent climate change – but, realistically, this is not likely to have enough effect. It is almost certainly a matter of too little, too late. Even if we persuaded the western world to give up its love affair with the SUV and cheap flights, the economies of China and India are gearing up to rival the US, currently the biggest influencer of climate change. It has been argued that the only way to prevent climate change from passing through a tipping point after which warming will accelerate beyond our control is to reduce greenhouse gas emissions by 90 per cent by 2030. No politicians are suggesting cuts that will achieve anywhere near this level of reduction.

Accelerating trends

We don't have to reach that tipping point to see climate change accelerating. Already the trends are getting worse. As *New Scientist* magazine said in February 2007, 'The [IPCC] authors acknowledge that they were being conservative. There is, though, a fine line between being conservative and being misleading, and on occasion this summary crosses the line. It omits some real risks either because we have not pinned down their full scale or because we do not yet know how likely they are.' Every week brings new revelations that global warming will hit us harder and sooner than was previously thought. John P. Holdren, president of the American Association for the Advancement of Science, wrote in 2007: 'Since 2001, there has been a torrent of new scientific evidence on the magnitude, human origins and growing impacts of the climatic changes that are under way. In overwhelming proportions, this evidence has been in the direction of showing faster change [and] more danger . . .' The world is on the brink of disaster.

Apart from a relatively small impact from the heat of the Earth's core, the world's warmth comes from the Sun. Without the energy of sunlight, the surface of the Earth would be similar to that of one of the distant planets in the solar system with a temperature hovering below the −250°C mark. The Sun's warmth is essential to preserve life – but it is also the Sun that pushes us into global warming. Normally a fair amount of the Sun's energy is reflected back off the Earth out into space. The more of that energy that is absorbed rather than reflected, the more Earth temperatures will rise.

Living in a greenhouse

The greenhouse effect, which we've heard so much about, modifies the amount of the Sun's energy that escapes the atmosphere. Again, like the Sun, this isn't a bad thing in itself. If there were no greenhouse effect, the Earth would be an unpleasantly chilly place, with average temperatures of −18°C, around 33 degrees colder than it actually is. But living in a gaseous greenhouse can be equally troublesome. The greenhouse effect is caused by water vapour and gases like carbon dioxide and methane in the atmosphere. Most of the incoming sunlight powers straight through, but when the energy heads back into space as infra-red radiation, some of it is absorbed by the gas molecules in the atmosphere. Almost immediately the molecules release the energy again. A portion continues off to space, but the rest returns to Earth, further warming the surface.

We only have to look into the sky at dusk or dawn when the planet Venus is in sight to see the result of a truly out-of-control greenhouse effect. Venus is swathed in so much carbon dioxide (around 97 per cent of its atmosphere) that relatively little energy gets out. Admittedly our sister planet is closer to the Sun than is the Earth, but it's this exaggerated greenhouse effect that has led to average surface temperatures of 480°C – hot enough for lead to run liquid – and maximum temperatures of around 600°C, making it the hottest planet in the solar system.

No one is suggesting that the Earth's atmosphere is heading for Venus-like saturation of greenhouse gases, but there is no doubt that the concentration of carbon dioxide, methane and other gases that act as a thermal blanket is going up. Each year we pour around 26 billion tonnes of carbon dioxide (CO_2) into the atmosphere. Around a quarter of the CO_2 we produce is absorbed by the sea (though this process seems to be slowing down as the oceans become more acidic), and about a quarter by the land (much of it eaten up by vegetation), but the rest is added to that greenhouse layer. Looking back over time – this is possible thanks to analysis of bubbles trapped in ancient ice cores from Antarctica and Greenland – the carbon dioxide level was roughly stable for around 800 years until the start of the Industrial Revolution. Since then it has been rising, and even the rate at which it rises is on the increase – the level of CO_2 in the atmosphere is not just growing, it's accelerating.

In pre-industrial times, the amount of carbon dioxide in the atmosphere was around 280 ppm (parts per million). By 2005 it had reached 380 ppm, and was higher than it had been at any time in the last 420,000 years. It's thought that the last time there was a consistent comparable level was 3.5 million years ago in the warm period in the middle of the Pliocene epoch, well before the emergence of *Homo sapiens*, and it's likely that levels haven't been much higher since the Eocene epoch, 50 million years ago. The IPCC predicts that if we don't change the amount of CO_2 we generate, levels could be as high as 650 to 1,000 ppm by the end of the century. The Goddard Institute for Space Studies (GISS) model, one of the best computer simulations of the Earth's climate, which reflects the impact of these changes on water patterns, predicts that most of continental USA will suffer regular severe droughts well before then.

The era of drought

By the end of the century, current predictions are that the tropics will live through droughts thirteen times as often as they do now. Drought is already on the increase. A 2005 report from the US National Center for Atmospheric Research notes that the percentage of land areas undergoing serious drought had doubled since the 1970s. South Western Australia, for instance, is facing a steady reduction in rainfall, leading to both potential drought and increased chances of bush fires.

As drought conditions spread, availability of water becomes restricted. Significant decreases in water output from rivers and aquifers are likely in Australia, most of South America and Europe, India, Africa and the Middle East. Countries like the UK are likely to get significantly drier summers, though they will be accompanied by stormier winters, and when the summer *is* wet, it will be wetter than usual. Across the world, drought will be dramatic. The 2007 report of the UN Intergovernmental Panel on Climate Change predicted that by the last quarter of the century between 1.1 and 3.2 billion people will be suffering from water scarcity problems.

Most historical droughts have been relatively short-term. Showing as statistical blips in the climate rather than marked permanent change, they did indeed cause devastation and disaster, but could be recovered from. A long-term drought provides no way out. Where these happen, civilizations simply disappear. After three or four years, the inhabitants of the drought area are faced with a simple choice of evacuation or death. A couple of years later and you have an abandoned region, littered with ghost towns and dead villages. Drought is no minor inconvenience.

Even where there is no immediate drought, a rise in tempera-ture can push previously lush areas into decline. Many parts of the world that are currently tropical forest – the Amazon rainforest has to be the best-known example – are predicted to change to savannah, grassland or even desert as carbon dioxide levels rise and a combination of lack of water and wildfire destroys the woodland. The Amazon basin, long touted as the lungs of the world, has already become an overall source of carbon dioxide, pumping over 200 million tonnes of carbon from forest fires into the air – more than is absorbed by the growing forest.

If things continue the way they are, the expectation is that the Amazon rainforest will be just a memory by the end of the century.

200 million tonnes of carbon from forest fires into the air – more than is absorbed by the growing forest.

This change from carbon sink – a mechanism to eat up carbon dioxide from the air – to carbon source is not just a feature of tropical forests. In 2005, scientists in the UK reported that soil in England and Wales had switched from being a carbon sink to a carbon emitter. As average temperatures rise, the bacteria in the soil become more active, giving off more CO_2. Remarkably, in 2005 this was already proving enough of a carbon source to cancel out all the benefits from reductions in emissions that the UK had made since 1990.

Positive feedback

A combination of decrease in rainfall over areas like the Amazon rainforest with increase in temperature is expected to result in a massive die-back. There is a similar expectation that temperate and coniferous forests in Europe and parts of North America will be drastically reduced. The picture isn't uniformly gloomy – there is some expectation of a northern expansion of forest in North America and Asia – but even so, the overall effect is that vegetation that has been soaking up carbon will, in our lifetimes, become the opposite, an overall source of carbon, kicking the greenhouse effect into positive feedback.

The best-known example of positive feedback is the howl from a loudspeaker when a microphone is brought too close. Tiny ambient sounds are picked up by the microphone, come out of the speaker at higher volume, are collected again by the microphone and are reamplified, getting louder and louder until they become an ear-piercing screech. One of the most worrying aspects of climate change is that the global climate also features a number of positive feedback systems, where a change reinforces the cause of the change, making the change happen faster, which further reinforces the cause, and so on. Positive feedback has often been omitted from predictions. As *New Scientist* put it in February 2007: 'The rising tide of concern among researchers about positive feedbacks in the climate system is not reflected in the [IPCC report] summary ... One clear need is to get to grips with the feared positive feedbacks.'

It's not just the Amazon rainforest and the Australian bush that are tipping into positive feedback, adding to the greenhouse effect. Other forests around the world, hitherto carbon sinks, are disappearing as temperatures rise. For example a combination of the increased temperature and the spread of pests is having a devastating effect on some Canadian forests. In one year, British Columbia lost 100,000 square kilometres of pine trees (three quarters the land area of England) to forest fires and disease. The local government estimates that 80 per cent of the area's pines will be gone by 2013.

Wildfires, destroying thousands of hectares of land and properties, are becoming increasingly common. In 1998 fires destroyed 485,000 acres (190,000 hectares) in Florida and 2.2 million acres in Nicaragua. Spain lost more than 1.2 million acres of forest to wildfires in 1994, while Greece and Italy each lost over a third of a million acres in 1998. Even in previously temperate areas like the UK, wildfires now pose a threat.

Agriculture will be forced to undergo major change. Traditional crops of hot countries – olives, maize, sunflowers – will take over in areas like southern England, while regions already growing such crops will find it increasingly hard to provide food. The 2007 IPCC report that forecast huge water shortages also predicted that as the twenty-first century progresses, up to 600 million extra people will go hungry as a direct result of climate change. If things get too drastic, perhaps our only hope will be a 'Noah's ark' of food – the vault being built by the Global Crop Diversity Trust in the permafrost of the Svalbard archipelago near the North Pole which will contain three million batches of seeds from all current known varieties of crop as a defence against the impact of global catastrophe.

Disappearing permafrost

There is an even more insidious effect of global warming that provides another positive feedback loop in the climate system – the melting of the Siberian permafrost.

In West Siberia lies a huge peat bog, around a million square kilometres in area (the size of France and Germany put together). Peat, the partly decayed remains of ancient moss and vegetation, is a rich source of methane, a gas that has around twenty-three times as powerful a greenhouse effect as carbon dioxide. The methane from the bog is frozen in place by the permafrost – a solid ice/peat mix that never melts. At least, had never melted until now. That permafrost is liquefying, discharging a huge quantity of methane into the atmosphere. By 2005 it was estimated that the bog was releasing 100,000 tonnes of methane a day. That has more warming effect than the entire man-made contribution of the United States. And thanks to positive feedback, the more the bog releases methane the faster it warms up, releasing even more.

The urban heat island

The impact of increasing temperatures is even worse for our city dwellers than the rest of the population, thanks to the heat island effect. In a normal environment, summertime temperatures are kept under control by night-time cooling. With no energy from the Sun hitting the dark side of the Earth, the parts of the planet in darkness can only lose heat, and where there are clear skies this can happen surprisingly quickly – witness the biting cold nights of the desert. But something goes wrong with this natural cooling process in a city. The pavements and canyon-like streets act as storage heaters, absorbing energy during the day that will keep temperatures relatively high at night.

This is the reason that many of the casualties of the European heatwave of 2003 were in cities. It's not a sudden, short snap of heat that is a large-scale killer, it's sustained heat that goes on day after day, and particularly heat that carries on through the hours of darkness. In the 2003 heatwave, it never got cool enough at night for relief. On 12 August 2003, Paris suffered a night-time temperature that never went below 25.5°C – stifling for the majority of city centre households without air-conditioning. Thousands died from the impact of the relentless heat held in place by the city streets. The final European death toll was over 35,000 from the heat and up to 15,000 more from the pollution that builds up, particularly over cities, in the warm still air.

Europe isn't alone in suffering the impact of sustained heat. Even though air-conditioning is much more widespread in the US, hundreds died in Chicago in July 1995 when hit by a heatwave of such sustained ferocity that on two successive nights the thermometer never dropped below 27 and 29°C respectively. To make matters still worse, warm air rises. The temperature difference between the ground floor and the top floor of a building can be enough to make the difference between comfort and trying to sleep in a virtual oven. Older high-rise buildings without air-conditioning but with relatively good air flow are particularly likely to roast inhabitants of their upper storeys.

The urban heat island effect is real; but it is factored out of climate-change calculations to avoid confusing the impact from greenhouse gases, which means that cities are likely to fare significantly worse than the predictions of temperature rise given by the climate-change models. It has been shown that urban heat islands don't contribute particularly to the overall warming of the planet (this can be seen because there is no link between changes in global temperature and average wind levels in cities, yet the heat island effect only arises on still days) but that really doesn't matter to the person in the city apartment. She will still suffer more than the models predict.

The dying conveyor

That's just how things are now. Heatwaves like that of 2003, which currently might be expected every twenty years or so, are likely to be annual occurrences by the end of the century according to our best predictions. In fact, it is quite likely on today's forecasts that such a

Heatwaves like that of 2003 could be more typical of the coldest summer of the decade by the 2060s.

summer would be average by 2040 and could be more typical of the coldest summer of the decade by the 2060s. The one mitigating factor that might benefit areas like the British Isles is the slowing down of the thermohaline circulation, the complex system of ocean currents that transports large amounts of heat from the tropics to northern latitudes. The section of this ocean conveyor system that runs in the surface layers of the Atlantic, preventing countries in Northern Europe from being more like Siberia in temperature, keeping Florida warm in the winter, and boosting temperatures on the Atlantic seaboard of North America, is the Gulf Stream.

There is some evidence that climate change will produce a reduction of strength in this ocean conveyor, largely because of the impact of fresh water from melting ice sheets. The collapse of the conveyor was the scenario dramatized in the movie *The Day After Tomorrow*, but this hugely overemphasized both the speed of the change and its impact. Early attempts to model the impact of climate change on the conveyor suggested that it might shut down entirely over this century, but current best estimates predict a decrease in strength of around 25 per cent. This will help mitigate the heat impact of climate change in the areas warmed by the Gulf Stream, but will not totally counter it.

Sea-level rise

As the planet warms up, the delicate balance of coastal life will be devastated. Sea-level rises go hand in hand with increasing temperatures. This is not just a case of irritating a few coastal sea creatures. Many of the world's great cities, from New York to London, and major sections of low-lying countries like Bangladesh are at risk of destruction by relatively slight increases in sea level. In the storm surge of 1998, 65 per cent of Bangladesh was inundated. It would not take much of a rise to make this a common occurrence.

Climate change has a double impact on sea level. The headline-grabbing cause is the melting of vast tracts of ice, increasing the volume of water in the sea, but there is a more direct effect too. As liquids get warmer they expand, and given the huge volume of water in the oceans the effect is far from trivial. Just a few degrees' increase in temperature is enough to push the sea level up around a metre from expansion alone. But the melting ice isn't only featured more often in the news because it looks more dramatic on the TV screen. Though initially expansion will be responsible for more rise than melting, the situation with the world's frozen supplies of water is heading for potential catastrophe.

A very visual illustration of the impact of climate change is the way that ice is disappearing from the North Pole in the summer. Not only is this happening on a large scale, but also Arctic ice is melting much faster than was expected only a few years ago. NASA satellites have revealed that in summer 2005, 730,000 square kilometres of ice that is normally permanently frozen melted: this is without historical precedent. The area of ice at its summer low was 20 per cent smaller in 2005 than in 1978. At least once in the last few years, the North Pole itself has disappeared entirely. This can happen because the Arctic isn't a land mass but a floating sheet of ice.

The good news here is that melting Arctic ice doesn't contribute to an increase in sea level. Floating ice is already displacing water just as a ship does – if the floating ice melts, the overall water level doesn't rise. But that doesn't mean the disappearing Arctic summer ice is a good thing. Not only is it a disaster for wildlife like the polar bear, it has a direct impact on global warming. Melting ice drives another of the positive feedback loops that are rife in the climate-change world.

As we've seen, it's the Sun's energy that heats up the world. But not everywhere is equal when it comes to solar warming. The lighter in shade a surface is, the more energy is reflected back out to space (greenhouse gases permitting). The glittering whiteness of an ice sheet is ideal at flashing back a good portion of the energy while the dark waters of the ocean absorb significantly more; water takes in more heat than does ice. So the more the Arctic melts, the more energy is absorbed, melting even more ice – positive feedback. And even though the melting Arctic doesn't contribute directly to sea-level rise, because of this positive feedback it does contribute further to global warming (and hence indirectly pushes up water levels).

Uncovering Greenland

Much more worrying than the Arctic from the point of view of ocean-level rise is Greenland. If we think of Greenland at all, it tends to be either as a cold little place between Europe and America, or the first example of dodgy advertising, when it was optimistically given a name that implies verdant pastures in an attempt to attract gullible Norse settlers. But in climate terms, the interesting (and potentially frightening) thing about Greenland is its ice sheet. More accurately, this is no mere ice sheet, it's an ice mountain range. The Greenland ice sheet covers over 1.3 million square kilometres (think France and Spain combined), and is mostly over 2,000 metres thick. Compare this with the United Kingdom's highest mountain, Ben Nevis, which only rises 1,300 metres above sea level. At its thickest, the ice sheet is 3,000 metres deep, over half the height of Canada's Mount Logan.

According to NASA, through the 1990s the ice sheet was shrinking by around 50 billion cubic metres a year. That's a lot of ice – but it would still take between 1,000 and 10,000 years for the Greenland ice sheet to melt completely. There's no room for sighs of relief, though. As Jim Hansen, director of the GISS and George Bush's top in-house climate modeller, graphically put it, '[Greenland's ice is] on a slippery slope to hell.' By 2000, the rate the ice sheet was melting had accelerated so much that it was already losing vastly more than had been estimated just ten years before. The assumption had been that the ice would gradually melt from the surface downwards, trickling its way to the sea as run-off water. But what is actually happening is startlingly different.

Lakes of water are forming on top of the ice sheets. These sheets aren't always uniformly solid. If there's a crack in the ice below a lake, the water can rush down, opening up the crevasse further as it flows until it has passed through the entire sheet to the bottom, where the water flow can eat away from beneath, enabling huge chunks of the ice sheet to float off the land. '[If] the water goes down the crack,' says Richard Alley of Pennsylvania State University, 'it doesn't take 10,000 years [to reach the base of the ice sheet], it takes 10 seconds.' And this is without considering the impact of the melting of the Antarctic ice cap, which is also on land, so contributes to sea-level rise.

If the entire Greenland ice sheet were to end up in the ocean, the extra water would raise sea level by 7 metres.

As if the disappearing ice sheets weren't enough, there is plenty of evidence that the glaciers around the world are also disappearing with unprecedented speed. Not only do these contribute to sea-level rise (the glaciers of Tajikistan alone hold 845 cubic *kilometres* of water), but water from glaciers is essential for the irrigation and drinking water of many countries. Around 10 per cent of north-west China's water supply comes from glacier melt water, for instance, and there are higher percentages elsewhere. Loss of glaciers will have a devastating effect on the economy and social well-being of a number of countries.

The rising tide

Sea-level rises are real and are happening. The Carteret Islands in the South Pacific are already being abandoned, their 2,000 inhabitants displaced by the rising ocean. The current best guess suggests the islands will be totally submerged by 2015. Perhaps even more striking is the fate of Tuvalu, another collection of islands in the South Pacific, which forms a nation in its own right. The 10,000 people of Tuvalu are also having to give up their homeland. Before long that country will be a small, modern-day equivalent of the mythical Atlantis, disappearing under the waters.

Many of the world's great cities are on a coastline and would have to be abandoned if sea-level rises reach a fraction of the 5 metres that now seems entirely feasible. The timescale for this is uncertain. Conservative estimates put the rise by 2100 at 0.5 metres, but this doesn't allow for the impact of positive feedback and the unexpected behaviour of the Greenland ice sheet. The rate of change in the Arctic perennial ice last year was eighteen times faster than was predicted just ten years ago. By February 2007, sea level

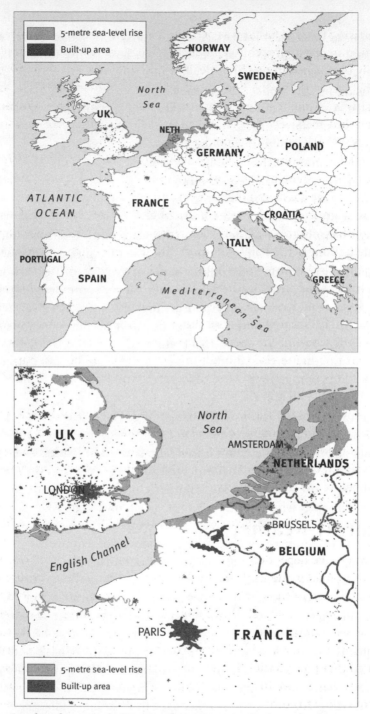

Land under water in Europe after a 5-metre rise in sea level.

was rising twice as fast as was predicted in 2001. Without a transformation in our approach to climate change, the 5-metre mark could easily be reached in our lifetime. It would only take an extra 3°C to bring the world to the conditions of the mid Pliocene, when sea level was 25 metres higher than today. Imagine the New Orleans flood, but massively deeper and never abating. Cities like London and New York would not stand a chance.

Global warming will change the shape of the inhabited world. Over 20 per cent of the world's population lives within 30 kilometres of the coast, and the number of people living in these at-risk areas is growing at twice the average global rate. Rising seas will mean that most of the US eastern seaboard would have to be abandoned, along with half of Florida, as will low-lying shore areas inhabited by hundreds of millions around the world. And this is not the limit. As we've seen, if the Greenland ice sheet melted entirely, sea level would rise 7 metres. The collapse of the fragile West Antarctic ice sheet would raise the level by up to another 6 metres, while the whole of the Antarctic ice cap melting would bring about an extra 60-metre rise (though this is thought unlikely to happen with temperature rises of less than around 20°C, so it's not one to hold your breath about).

Any figures for sea-level rise also need to be topped up with the impact of storm surges. In some areas – around the south-east of England, for example – when the storms are at their height sea level is expected to rise around a metre more than is otherwise predicted, well before the end of the century.

Stormy weather

We might be dependent on energy coming in from the Sun, but in terms of matter, the Earth is largely a closed system. Extra droughts in some parts of the world mean more wetness elsewhere. As well as the impact of sea-level rise, some parts of the world can expect increased rainfall, and particularly more heavy storm rain. At the moment the increase is relatively slight – in 2001, the IPCC estimated that precipitation in the northern hemisphere had increased by between 5 and 10 per cent over the previous 100 years – but there's more to come.

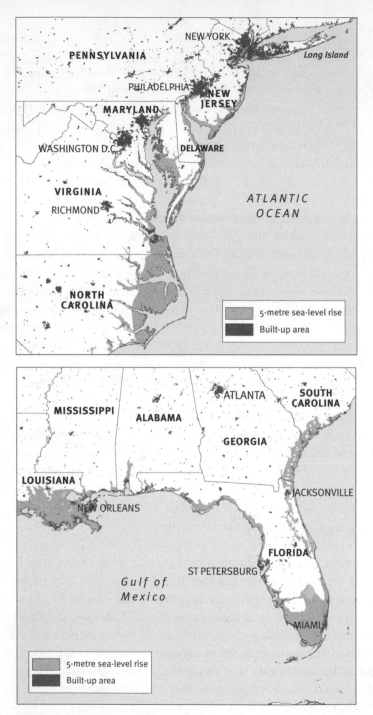

Land under water in The US (Gulf Coast and New York area) after a 5-metre rise in sea level.

Land under water in Australia after a 5-metre rise in sea level (note most of the occupied land in Australia is on the coast).

A significant fear is that global warming will produce more hurricanes like Katrina, the storm that devastated New Orleans and the surrounding coast in 2005. There is no certain evidence that climate change was behind the significant rise in *numbers* of hurricanes in 2005. As the oceans heat up, it should be easier for hurricanes to form, but there are other factors that come into play, and scientists are reluctant to commit themselves to saying that hurricane formation is on the increase. (This is a reassuring counter to those who think climate-change scientists have a hidden agenda and make predictions that show that man-made climate change is responsible for everything that goes wrong with the weather.)

However, even if the higher number of hurricanes in 2005 was a blip, it does seem true that they are increasing in power. Two studies in 2005 both showed that the energy levels of hurricanes is on the rise, with twice as many storms at the highest category 4 and 5 levels as were recorded in the early 1970s. There shouldn't be a similar effect with powerful tornadoes, though. The really big tornadoes are the product of a very special kind of thunderstorm that doesn't seem to be influenced by global warming. The smaller, more common tornadoes like those experienced in Europe may be on the increase, but equally it could be that we are just noticing and reporting them more.

The new disease threat

Changes in climate are not only affecting sea level, but also spreading disease. Malaria was once rife in the south-east of England. It would take only a few degrees' rise in temperature and disease-carrying mosquitoes could set up permanent home across much of Europe and North America. Today, malaria alone already kills a million people a year. Along with dengue fever and West Nile virus, it is increasingly turning up in locations that were once thought beyond its reach.

Global travel means that disease-carrying insects regularly hitch a lift on aircraft or in passengers' luggage, reaching far beyond the normal travel range of the insects themselves. Up to now the climate they faced on arrival was too much of a challenge, but there is good evidence that a changing world is becoming more accommodating to these tiny terrors. West Nile virus, for example, got a toehold in New York in 1999. By 2005 it had spread to cover much of the United States and southern Canada and was already responsible for the deaths of over 800 North Americans. (The virus, which largely resides in birds, seems to have spread more widely among humans in America than in Europe, because whereas in Europe the mosquito mainly responsible for spreading the virus comes in two types, one specializing in birds, the other preferring people, in the US and Canada a hybrid has developed that is equally fond of both. This could easily spread to Europe, but hasn't yet.)

Studies at Durham University show that malaria was wiped out in England by a series of cold summers in the 1800s. The climate conditions are now heading in the right direction – increased precipitation and warming temperatures – to make malaria's return to previously infection-free regions like Western Europe and much of the continental United States a likelihood.

It's not just people who are at risk from climate change opening up new territory to parasites. The Colorado beetle, which has devastated American potato crops in the past, has rarely gained much of a foothold in Europe because the temperature hasn't been high enough for it to thrive. But just a couple of degrees of temperature increase – inevitable, according to nearly all predictions – will be enough for the beetle, which is already making predations into Canada, to be comfortable eating its way through practically any potato crop in the United Kingdom and other European countries at a similar latitude.

Losing essential services

As things get worse, there will be huge disruption to normal services. Availability of electricity, petrol and gas will be increasingly restricted as the need to respond to climate change goes critical. At the same time, with stocks of non-renewable fuels running short and sources of supply becoming more remote, there is a growing opportunity for disruption of supply by natural disasters and terrorists. We could see a regular or even permanent breakdown of services that are essential for our everyday lives.

Of course, prediction isn't an exact science. We can't even forecast the weather more than a few days out, so it seems optimistic to assume that we can know how the world's climate will change over tens of years. But while there will always be varying interpretations as long as there are different scientists analysing the data, the consensus is now hugely in favour of global warming being a real, growing threat.

Some sceptics still point out that the changes to date are relatively slight, and may not have a huge amount of impact before the end of the century, but they are missing two significant points. First, the impact has begun. If you doubt this, speak to a citizen of New Orleans, someone who has lost a family member to the heatwave that caused thousands of deaths in Europe in 2003, or someone who has lost their home to unprecedented wildfire or coastal floods. Secondly, it's a mistake to assume that the relatively slow rate of change we see now will continue at such a creeping pace.

The sudden swing

According to Will Steffen, the Australian climate-change expert, the world is not usually a place of gentle, slow drift. 'Abrupt change seems to be the norm, not the exception,' says Steffen. On twenty-three occasions during the last ice age, for instance, air temperatures went through massive climbs, pushing temperatures in Greenland up by as much as 16°C in about forty years. Around half of the rise in temperature that led from each ice age to the following inter-glacial periods – again, changes of the order of a huge 16°C – took place in just ten years.

When the Earth undergoes major change it tends to be in sudden, large steps – this is something that is a relatively recent discovery. Richard Alley, in a report for the US National Academy of Sciences, concluded: 'Recent scientific evidence shows that major and wide-spread climate changes have occurred with startling speed . . . this new thinking is little known and scarcely appreciated in the wider community of natural and social scientists and policymakers.' We might be predicting a half-metre rise in ocean level by 2100 (plus up to another metre of storm surges) based on current, slow, steady rise, but we have to prepare for the possibility of a precipitous step-change in temperature that will result in much faster rises in sea level.

Even without step-change, there is a real possibility that current predictions are underestimating the impact, because they have not taken sufficient account of positive feedback accelerating the process of global warming.

Cleaner air means worse warming

Sometimes, even our attempts to make the environment better can have an ironic and unexpected effect. Aerosols – the scientific term for suspensions of fine particles in the air, typical of much airborne pollution from smog to black smoke – have been cut back significantly as we manage to clean up the air. But aerosols have a helpful effect where global warming is concerned. Unlike greenhouse gases, they stop the Sun's energy on the way in, so have a cooling effect on the ground below. (This is reversed if soot particles, for instance, land from the aerosol on snow, darkening it and reducing reflection.) At the moment aerosols could be reducing the global warming impact of greenhouse gases by up to half – but this contribution is liable to seep away as we achieve cleaner air.

Human or natural?

A handful of scientists question whether human intervention is responsible for global warming. They point to pre-human variations in temperature, which clearly weren't caused by our production of greenhouse gases, and ask why global temperatures fell during the period between the 1940s and the 1970s, when the western world was going into industrial overdrive. Our current warming, they suggest, is down to solar activity, either directly, as a result of variations in the Sun's output, or indirectly by a reduction in the impact of cosmic rays on the Earth, which in turn reduces low cloud formation, letting temperatures rise. They also point out that in pre-human times, global warming tended to precede a rise in carbon dioxide level, rather than the other way round.

Mainstream scientists don't doubt that the Sun and cosmic rays have some input on climate change, but the vast majority are now convinced that the main factor in our current warming is human-produced greenhouse gases. It is true, they say, that pre-human warming events began before carbon dioxide levels rose. The warming released trapped carbon dioxide, which then resulted in further warming and so on. But there is no doubt whatsoever that carbon dioxide is a greenhouse gas that produces warming effects, nor that human activity is increasing CO_2 levels in the atmosphere. As for the cooling between the 1940s and the 1970s, this is to be expected as an effect of the pollution emitted in that period, blocking sunlight with aerosols of fine particles. The best scientific evidence suggests that the sceptics are wrong – but it doesn't really matter from the point of view of surviving the impact of climate change.

Even the sceptics accept that global warming exists.

Everyone also agrees that there's a lot of uncertainty in the predictions of just how great the effect will be. That's inevitable because they are dealing with a very complex and only partly understood system. Clouds, for example, have a big impact on the climate. Low clouds have a cooling effect; high clouds trap infra-red radiation and warm us up. Different types of cloud make very different contributions to heating or cooling. Attempting to include the feedback produced by clouds into climate models produces a huge range of variation.

This means that though the predictions held to be most likely are still for a temperature rise of 2°C–3°C within a century, it certainly isn't impossible that it might be 10°C or even 12°C – plenty to make the most dire predictions of the impact of climate change a reality in our lifetime. A stark indication that this may be the case came from a 2007 report by the US Academy of Sciences, stating that climate change indicators have recently been rising three times faster than the worst predictions. The knowledge that there's uncertainty doesn't mean we can just cross our fingers and hope the threat goes away. It's all the more reason to be ready, in case things head for the unpleasant end of the prediction range. And something we know for certain is that whichever of the predicted averages proves correct, we will experience worse than that.

The projections we see for the impact of climate change are based on averages, rather than extremes, but we don't experience averages. Compare a weather forecast for tomorrow with the average picture of the weather in your area at this time of year. It's the forecast that tells you how the weather will affect you, not the average. We want to know what the storm surges, the heatwaves and the hurricanes will do to us, not how things average out. The damage is caused by the worst the weather can throw at us.

Being prepared for global warming means thinking through the possible consequences for you, and putting some plans in place. That's what *The Global Warming Survival Kit* is here to help with. If we're lucky, and disaster doesn't strike, it will still come in useful if there's a power cut or your water supply is temporarily cut off. If we're not lucky, it's going to be a life-saver.

TWO

POWERLESS

We are hugely dependent on electricity, gas and other fuels. Blackouts and fuel shortages can have a drastic effect on our lives. Need some immediate help? This section includes:

What to do ...

The cocoon of power

We live a fragile life, protected by a cocoon of power. The ready availability of electricity, gas and petrol, the source of this power, is something we take for granted. Price rises may cause irritation – but the power is always there. This can't be assumed for much longer. Power is becoming scarcer. It won't always be this easy.

Oil supplies have been occasionally disrupted by events in the Middle East for many years past. As we bring more of our energy sources from a greater distance, so the risk increases. Countries which historically relied on their own gas and coal have now to import more and more from further away. Even the United States already imports 15 per cent of its gas; the UK is expected to import 50 per cent of its needs by 2010, and some Baltic states rely entirely on foreign sources. Massive power conduits like gas pipelines make easy and attractive targets for terrorists – and are highly susceptible to natural disasters.

As more resources are concentrated through fewer routes, they also become at risk from political and commercial upheaval. When in 2006 Russia turned off its supply of gas to Ukraine in an argument over pricing, Serbia also lost half its gas, with Hungary, Croatia and Slovakia all losing a sizeable portion. As western countries take more gas from further away, this scenario will happen elsewhere. Any disruption to the gas supply also has an impact on electricity, as many countries have moved the generating balance away from coal to cleaner gas-fired power stations.

At the same time, as global warming begins to bite, there will be sudden and large-scale government responses, restricting access to power. Although there is increasing pressure for clean generation of energy, many countries derive much of their electricity needs from coal- and gas-fired power stations which make large contributions to climate change. When in 2006 protesters attempted to shut down the UK's Didcot coal-fired power station they were complaining both about the huge inefficiency of the plant – around two-thirds of the energy generated is lost as heat to the atmosphere – and its carbon dioxide production, which at around six million tonnes a year was comparable at the time to the total combined output of the twenty-nine countries lowest on the carbon-emitting scale. As governments bite the bullet to reduce emissions, power supplies will suffer. There just isn't time to develop renewable sources on a sufficiently large scale, or to build nuclear power stations (which also have low climate-change impact) before supplies begin to suffer.

Climate change also contributes directly to power outages. Extreme summer temperatures often lead to failures of electricity supply: the increased demand from air-conditioning can overload the system; and the heat makes power lines expand and sag, causing blackouts if they come into contact with nearby trees. The dramatic weather systems generated by global warming, including tsunamis and hurricanes, can also wreak havoc with power distribution. In December 2006 around 1.5 million homes in the US states of Washington and Oregon were blacked out, some for up to a week, as power lines were brought down by howling windstorms and heavy rains. In January 2007, storms in the UK left 300,000 households without power, many for several days. As the impact of climate change grows, these will become very familiar headlines.

It doesn't help that the electricity grids of many countries are suffering from overload and age. Much of the UK's National Grid, for example, needs replacing before 2020 if it is not to fall into disrepair. As systems become more complex, their susceptibility to freak accidents and technical problems grows. In 2003 there were two large-scale electricity blackouts in western countries. The Northeast blackout affected a sizeable part of Canada and the north-eastern US, leaving a total of 50 million people without power. The result was not only loss of electricity, but disrupted water supplies for millions who rely on pumped water, chaotic disruption of transport, massive losses for business, restrictions on medical capacity, limited communications and looting. The same year, an enormous power outage left the whole of Italy and parts of Switzerland without electricity, causing upheaval for a record 56 million people.

Rural areas regularly experience power cuts already, whether from lightning strikes or system failures. Blackouts and shortages will become an increasingly familiar experience for everyone, resulting in sudden loss of power for days or weeks. No light, no TV. No cooking and no heating. Try to ring the power company to see what's caused the power cut: no phones. No internet. No mobile network.

Welcome to the powerless life, a world that has been the norm for most of humanity's existence. We have become used to power at the push of a button. When it isn't there, the disruption is both shocking and immediate. Some shortages take time to work their way through a country. If food runs short, many of us have some stocks to keep us going for a while. Electricity is different – one moment it is there, the next it is gone. The first notice you get is when the power is no longer available.

As human beings we are particularly good at ignoring threats, even very obvious threats, if they don't give us warning of immediate disaster. As the situation becomes ever more serious, the majority of the population will take astonishingly little action to protect themselves, even in the face of the starkest possible warnings. This, unfortunately, is a widespread human characteristic. Some, however, will take heed and prepare accordingly. Forewarned is forearmed.

The inhabitants of Pompeii knew perfectly well that Vesuvius posed a mortal risk, but chose to hope for the best

Every day in hotels around the world, if you know where to look, you will see off-duty air crew perform a strange ritual before settling down to sleep. They come out of their rooms, find the nearest fire exit and memorize the route, counting doors along the way. It is part of their training to survive. A hotel fire is by definition an unanticipated event, one that is significantly more likely to claim lives than a plane crash. The unexpected nature of a hotel fire is exactly why it has the potential to be so deadly. A few wise folk will take sensible precautions to prepare for the unexpected, but most won't.

Exactly the same need for preparedness applies to power outages. It might seem unnecessary to do anything about it right now, just as walking the route to your nearest fire exit will probably seem too much trouble next time you stay in a hotel. But by the time you know it is necessary, it will be too late. Make sure that you are ready.

Solutions

Be prepared

You may not have any warning of a power cut – the sections below give advice on action to take when a blackout occurs – but if you have a chance to prepare, there is a lot that you can do to make a difference. Take a look through the checklists on pages 73–86. Stock up on the basic kit, and consider the options in the advanced kit like fitting a multi-fuel stove and ensuring that your insulation is up to scratch: these are more costly, but can transform your ability to keep your home comfortable.

The first minutes of a power cut

At home

The power shuts down. Suddenly most of the activity of modern life disappears from your home. It's a blackout. Take some simple steps to minimize the impact of the power outage:

❏ Calm any children or pets.

❏ If it's dark, get a light source – a torch, candle or, if these aren't available, illuminated electronics like a mobile phone.

❏ Check the mains supply for tripped circuit-breakers or blown fuses. This might not be a power cut.

❏ If you were working on a computer with battery back-up, save what you can. Shut down as quickly as possible.

❏ Keep your food safe. Don't open the fridge or the freezer more than you have to. A modern, well-insulated freezer, if full, can keep food frozen for up to 48 hours without power. If there's not much in the freezer, transfer cool bag blocks and ice to the fridge. The first time you have to open your freezer, quickly insert extra insulation (see page 51).

❏ Check your communications. Landline phones and some mobile networks have a separate power supply. Ring the power company. Ring friends to see how far the blackout spreads. Try a corded landline phone that doesn't use mains power first, then your mobile. If you don't have any working phones, try the nearest payphone.

❏ Turn off electrical equipment with electronics (pretty well everything electrical, including washing machines, tumble-dryers and toasters) at the socket, as it may be damaged if there is a brief 'spike' of electricity when the power is restored. But leave a light switched on to alert you to the power coming back.

❏ Deploy any power outage equipment (emergency lighting, heating, cooking etc.) you may have.

❏ Check local radio, then national radio, for any information on the outage. If you think you don't have a battery or wind-up radio, check for 'hidden' radios – a shower radio, or a radio in an MP3 player or phone, for example.

❏ If it looks as though the power cut may last for some time, at the first opportunity top up your emergency power sources and fill any gaps in your power outage equipment.

At work

A blackout in the workplace – particularly a high-rise building in a city – can be frightening. Take a moment to calm yourself. Unless there is an obvious threat, it is best to stay in place for a reasonable length of time in the hope that power will swiftly be restored, before you consider leaving the building.

□ Don't panic. Take some slow, deep breaths. Many buildings have emergency lighting which may take a few seconds to come on.

□ If it's dark, get a light source. Get the emergency torch from your bag/pocket/desk if you have one. Otherwise find something to make a light. Remember portable electronics (phone, PDA etc.) usually light up.

□ If you were working on a computer and have battery back-up, save what you can. Shut down as quickly as possible.

□ Check your communications. Is the phone working? Can you establish what has happened? Try phoning whoever is responsible for the premises or building management.

□ If it is necessary to leave the building, use the emergency exit and stairs. Don't stay on the streets longer than you need to – get home.

On the street

A blackout on a city street at night can be highly unnerving. Get your back against the wall of a building, to provide stability in case of panic; it's all too easy for someone in a crush to be pushed over, then trampled. Give yourself a few moments to get used to the lack of light. If you live within walking distance, get home as soon as you can. If you are driving, get to your car, but don't drive off yet – wait until drivers have got used to the conditions and have worked out how to deal with junctions. Cyclists may find it advisable to stay off the road initially, and should be sure they are wearing high-visibility clothing. Initially there may be a fair number of collisions. If you use public transport and can't walk home, head for the station or bus stop, but expect delays. Trains, whether underground (subway) or overground, are unlikely to be running. If you intended to use one, can you walk to your destination instead? Be particularly careful crossing roads at night – you are much less visible in car headlights alone than you are with street lighting, and the drivers will be disoriented.

In a car

If it's dark, turn on your headlights. It should be obvious, but some people will fail to remember this. Do not rely on side-lights – these are only intended for parking, and won't show up any hazards. Slow down and keep a vigilant eye out for pedestrians as they will be harder to see. Whether it's dark or light, junctions that were controlled by traffic signals will become hazards. Approach any junctions carefully: treat all traffic lights and other control signals as stop signs. If it's practical, get out of a built-up area – country roads have few powered features to become hazards.

Be particularly careful on railway crossings, which may not have power for their warning signs. Approach the crossing with care. Stop and look both ways down the tracks. Only proceed if there is no train in sight. A train travelling at 200 kilometres per hour (125 miles per hour) will cover 55 metres (183 feet) in a second. You need plenty of clearance.

In a strange building

If you are sitting down, stay seated. If standing, try to get your back to a wall. Wait for a while to get used to the lighting conditions if it is dark. Unless there is a pressing reason to get out (fire, for instance), stay in place. One of the easiest ways to get hurt in these circumstances is to be trampled in a rush to get out. Wait until it's clear that the power isn't going to be restored any time soon. Only leave once you have a clear picture in your mind of where you are going to go, and how you will get there with no power.

If the exit isn't visible, look out for green emergency exit signs – these should have battery back-up in the dark in case of power failure. Even if lifts appear to be working, don't use them – stick to the stairs. It is possible that any emergency power may last only a few minutes then cut out, leaving lifts stuck between floors.

On a train

On a subway or underground train, there may initially be complete darkness. Don't panic, and stay where you are. If the train is still moving, brace yourself as you will not get the usual indications of stopping. There should be some form of emergency lighting. Do not try to leave the train. This is extremely important – power could be restored at any time, and leaving the train could prove deadly.

Wait for a train company official to give instructions. If you need to evacuate the train, on a conventional line get as far away from the tracks as possible – up an embankment, for example. On a subway/underground train you will need to make your way along the tracks to the nearest station. If the train has powered rails (more than two rails per train), stay away from the rails, even though the power is off. It may be restored without warning, making contact with the live rail fatal.

In a lift

A lift is a slightly unnerving place at the best of times – you are forced into closer proximity to other people than is strictly comfortable and suspended over a deep shaft by a few cables, so this is hardly surprising – but it is significantly worse in a blackout.

First essential: don't panic. Even if there is a total loss of power, there are automatic brakes that will clamp on to the cables and hold you in place. The lift is not going to drop down the shaft. If the buttons are still illuminated, try pressing the button for the nearest floor or the ground floor (not every one in sight). Then try the alarm button or phone – these may have emergency power, even if the lift itself is not operating.

The most important thing is not to behave like an action hero and attempt to climb out of the lift itself into the shaft. There may not be a hatch, but if there is, it's not there to provide an escape route. There is a collection of dangerous machinery above the lift, you could easily fall, and there really is nowhere to go. To make matters worse, if there is a hatch, it will probably immobilize the lift if you open it, and if it doesn't, the lift might start moving while you are out there. This is not a sensible option. Realistically, you are unlikely to be able to get out of the shaft by forcing the doors open (apart from anything else, you probably aren't opposite the external doors).

Try the mobile phones of everyone in the lift. They might not work because of the metal around you, but they often do, and this could be the quickest way to get assistance. If the alarm and phone don't seem to do anything, bang on the doors and shout at regular intervals, leaving gaps to listen for a response. Again, don't panic. You aren't going to be dropped down the shaft, nor will the air run out in the lift. You will be rescued, but it can take some time. It's not easy to be patient in such circumstances, but do your best. Share stories with other trapped travellers.

Temperature control

Staying warm at home

If it is a cold day (or night), once the power goes the temperature will start dropping quickly. The better insulated your home, the slower this will be. Staying warm relies on two essentials – generating heat and keeping it in place with insulation. Each of these will help keep you and those around you safe if temperatures fall.

The smaller the space you have to heat, the less of your precious energy it will take. Consider keeping only one room warm. If you have a room with a self-contained heat source (an open fire, stove or Aga-style cooker, for example), use that. Otherwise pick the warmest room that is well insulated

– remember heat rises, so this may be one of the higher rooms in the house. Once you have chosen a room, there are a number of ways to build up its insulation:

❏ Insulate the windows: even with double glazing, these will lose heat faster than the walls. Draw any curtains or blinds. Stuff the space between the curtains and the windows with insulating material – foam polystyrene, loft insulation in plastic bags, or failing this, newspaper and cardboard.

❏ Make sure the heat loss through the ceiling is minimized (heat rises). If this is a long-term situation, move the loft insulation from the area above distant rooms to increase the insulation above the one you've chosen. (But leave the insulation above the surrounding rooms: if they stay fairly warm, they will act as a heat barrier around your warm room.) If you are using a downstairs room, put insulation on the floor above. As well as loft insulation in bags, polystyrene etc., you can use spare bedding and mattresses.

❏ Check doors and windows for draughts: block sources of cold air with a draught excluder, wadded clothing or loft insulation. If there is an external door in the room, treat it like a window, rigging up a curtain and insulating the space between curtain and door.

Consider your own insulation too. It might seem unnecessary to be wrapped up as if for outside when sitting at home, but it's better than getting cold. Use multiple layers to build up insulation, and don't feel silly if you need to pile on blankets or a duvet. If you have to, block the gaps in your clothing (for instance around collars and cuffs) with scarves or other insulating material.

If your water heating is out with the blackout, make sure your hot-water tank is well insulated. For the moment, if your tank isn't properly insulated, use what you have available (a duvet, polystyrene sheets, or loft insulation, for example). When you get a chance, obtain a proper insulating jacket for the tank, or have it replaced with one of the newer, pre-insulated tanks. Good insulation on the hot-water tank can reduce by around 25–45 per cent the energy required to heat the water in the first place (good for your bills) and will keep it at a useful temperature significantly longer after the power goes.

To avoid getting cold yourself, minimize the amount of your body that is in contact with something cold, or something that is a good conductor (like a metal) which will lead heat away from your body. Make sure a metal chair or equivalent is well covered with insulating material before sitting on it. The ideal to minimize heat loss is to be in contact with something at body temperature – that way there is very little flow of heat. Snuggling up with other people will reduce the amount of heat that is escaping from your body.

POWER TIP
Hats aren't as important as you've been told

The whole point of clothing for insulation is to stop heat loss from your body. You may have heard that 50 per cent (or even 75 per cent) of your body heat is lost through the head. This is a myth that originated in a sales campaign for hats, back when it was OK to tell lies to sell more stock. The actual figure is more like 10 per cent – so it is worth wearing a hat, but it really isn't going to make as much difference as you might expect. It is more important for babies, though: with proportionately bigger heads, they can lose 15–20 per cent of their body heat that way.

Heat generation is a matter of converting energy from other forms into heat. You can produce energy in a number of ways. With no materials, a small amount of heat can be generated from muscle action – for instance by rubbing your hands together, which produces heat from the friction – but most effective heat sources involve either burning something or converting an existing form of energy such as electricity into heat. Note that drinking alcohol may produce a 'warming' sensation, but does not heat you up – in fact a nip of brandy will encourage heat loss, not make you warmer. Avoid alcohol if you are cold.

The burning process is a chemical reaction turning one material into others. This often results in the production of gases, which may be toxic. Carbon monoxide, which can be produced from burning many materials, is a particular hazard: colourless, odourless and deadly. Make sure if burning something in a confined space that there is adequate ventilation, with an exit route that will lead the gases away (a flue, for instance). It's also important to make sure that fires are contained and don't spread. When making a fire indoors use a professionally made enclosure like a stove. Never use charcoal, for instance barbecue fuel, indoors – this produces a considerable quantity of toxic fumes.

Not many years ago, practically all houses had open fires, and because central heating was uncommon, many households would also have portable paraffin heaters or similar heat sources not dependent on the mains. If you have a fireplace that isn't in use, consider having it reinstated, or even better have a multi-fuel stove fitted into the fireplace, which will be able to provide heat in a more efficient, more controlled and safer fashion.

POWER TIP
Fuelling your stove

You may need to be creative about sources of fuel for your stove if you haven't enough stored. First check the house and garden for unwanted lumber. Stick to wood as much as is possible – paper does not burn very efficiently (the same goes for books, and it just feels . . . wrong) and if your fuel has bits of metal etc. in it you are liable to clog your stove up with debris. Avoid plastic, which can produce poisonous fumes. It's fine to chop up the dead tree in the garden, but be careful using old Christmas trees and other conifers – the high content of resinous sap will make them flare up, and tends to leave a layer of gooey tar on your stove glass. Use conifers in small amounts with other wood.

If you have access to fields or woodland, using old wood is fine (with the permission of the landowner) but don't cut down living trees. Remember garages and attics where you may have stocks of 'might come in useful' wood or old unwanted wooden furniture.

If you haven't had the chance to get a stove installed, or your fireplace and chimney professionally reinstated, take care, as it's easy to start an unwanted fire in the chimney or to have fumes pouring back into the building. Check the fireplace and chimney for blockages, and make sure there is still a heatproof firebrick at the back of the fireplace (these are sometimes removed when a fireplace is out of action). If you can't borrow proper chimney-sweeping brushes, put a small shrub like a compact holly bush on the chimney top (take the usual safety precautions in getting on to the roof), and drop an attached rope down the chimney, pulling the bush through to sweep. Be ready for a flood of soot in the room below. This can have very fine particles that may cause breathing difficulties – always wear protective clothing and a mask. When a fireplace has been reinstated, be wary of fumes leaking from the chimney breast further up the house, for example in a bedroom above.

If you haven't a fireplace or chimney, consider buying a portable heater for emergencies. Paraffin heaters are still available in some countries, but it's more common now to have one that runs on a liquefied gas such as propane or butane. If you do buy such a heater you should have it checked on an annual basis to ensure it is burning safely – the gas vendor or heater vendor should have information on how to do this.

If you use a portable heater, whatever the fuel, make sure it is in a safe location where it can't be easily knocked over – don't site it in a corridor or too close to a door. Never put clothing over this kind of heater to dry – this can easily start a fire. Store any spare gas cylinders in a well-ventilated location above ground (the gas is heavier than air, so will pool below ground). Don't leave portable heaters on overnight, or use them in small, poorly ventilated rooms.

With no electricity, gas central heating is liable to be useless, but if the gas supply is still on, a gas cooker is a potential source of heat that is designed for indoor use, though it shouldn't be used for more than a couple of hours at a time. Admittedly it isn't a very efficient way of providing heat, and you may well not want to live in your kitchen, but remember it's there if you have no better way of staying warm. However, never try using a patio heater, barbecue or camping stove as an indoor heater. The fumes they produce are extremely dangerous in confined spaces.

⚡ POWER TIP

Take care using camping equipment indoors

In the US alone, around 30 people die and 450 are injured each year, poisoned by using camping heaters, lanterns and stoves indoors and in tents. The gases used as fuel (propane and butane) aren't themselves poisonous, but they produce carbon monoxide when burned. Don't use camping equipment indoors. Use it out of doors, in a well-ventilated space. If weather conditions prevent this, in emergencies use only for cooking and as briefly as possible – as far away from the occupied room as possible and with the best possible ventilation. Also bear these points in mind:

☐ Use a carbon monoxide detector.

☐ Never use gas-powered camping equipment while asleep – it is easy to succumb to carbon monoxide poisoning in your sleep.

☐ Watch out for the symptoms: headaches, feeling dizzy and weak, nausea and vomiting. Carbon monoxide poisoning is often mistaken for a cold or flu: be wary if you or your family appear to be going down with an infection.

☐ Those most at risk are pregnant women, babies and young children, the elderly, smokers and those with blood or circulatory system conditions, including heart disease. Both alcohol and drugs make the effects of carbon monoxide poisoning worse.

Heat and cooking outdoors

If there is no practical way to provide heat and cooking power indoors, or if you haven't got access to the house for a period of time, you will have to resort to a traditional fire. Lighting a fire outdoors is harder than it looks. It's much better if you can practise this when you aren't under pressure, to get the hang of it before it's a necessity.

Use a cigarette lighter or waterproof 'lifeboat' matches, if you've got them, to start your fire – in many conditions ordinary matches will blow out before you get them to the kindling. Build your fire with a layer of twigs at the bottom to raise it off the damp ground, then starter kindling like crumpled newspaper or very dry fibrous plant material – dried bracken, for example. On top of this put thin kindling wood, building up through bigger sticks to your main sections of wood.

Always use wood that is as dry as possible – preferably not wood that has been sitting on the ground, rotting. Highly resinous wood like an old Christmas tree or other conifer is good to help in starting the fire, but burns up too quickly to be of use for heating and cooking (watch out for flare-ups).

If you don't have a source of a flame, you can make a spark from a flint and steel combination (an old dead cigarette lighter is a handy portable version), or friction (for example bowing a stick that's resting in a depression in a piece of wood containing very fine tinder) – but this is extremely difficult. It's better to have the right tools for the job.

Staying cool at home

It seems natural that lack of power will be a problem in cold weather, but power is also essential for staying cool when it's hot – and climate change is producing summer temperatures in more and more locations around the world that push human bodies to their limits. During the summer of 2003, a total of 35,000 people died across Europe from the direct impact of heat. As usual with extremes, everyone is susceptible, but the first to be affected (and thus to need special care) are the very young, the elderly and those suffering from illnesses that lower the body's ability to cope with exertion.

The actions you can take have the opposite intentions to those used to keep warm. You need to keep out the heat and to remove heat from the building. The main source of heat, and the cause of our heatwave problems, is the Sun. Keeping direct sunlight out of a room is an important part of staying cool, particularly when you have large windows or patio doors. If you have external sun blinds or shades, deploy them. Use any sun umbrellas, tilted at an angle, outside the window to reduce the amount of sunlight getting through. Draw the curtains. Make sure your drapes and blinds are of bright, reflective material – the more they reflect, the less that gets into your house. A longer-term possibility is to give your roof (and to a lesser extent, your walls) a light-coloured coating that is as reflective as possible.

If it's warmer outdoors than inside, *don't* open the windows – all this will do is heat up the room – but make sure you do have the windows open when it's cooler outside.

A pair of thermometers inside and out, shaded from the sun, can provide a useful guide. Stay in lower-storey rooms, which will be cooler than upper floors. Move your beds down to the lowest part of the house to make hot nights more bearable.

When a blackout hits, your fridge and freezer will stop working, but you can make use of your stockpiled ice packs to cool down (bear in mind the risk to your food; see the next section, 'Rescuing food'). There's really not a lot of point in just leaving the fridge door open – the amount of residual cool air from the fridge will be quickly overwhelmed by the volume of air in the house – but you could sit near it with the door open for a bit of relief if you have no food inside to keep chilled.

One of the most important aspects of coping with heat is countering dehydration. When we're hot, we sweat more, so that evaporation can cool our skin. It's easy for the body to get low on water, and that's where heat-related health problems really kick in. Drink water sensibly. This means taking regular sips when you feel thirsty, not pouring down litres at a time. In excess, water can over-dilute the sodium levels in the body, causing disorientation and even death. Cool baths, or a flannel soaked in cold water, can also be used to relieve the unpleasant sensations of being too warm.

Rescuing food

Fridges and freezers will begin to warm up as soon as the power goes off. Don't open the doors more than you have to. With the power off, a fridge or freezer is effectively an insulated box – the more you open the doors, the more heat you let in. Make sure everyone in the family knows this. If your freezer is well packed, don't open the door until you have to – at that point, quickly insert sheets of foam polystyrene or other insulating resources around the inner walls, then close up as quickly as you can. Unopened, a full freezer can keep food safely for 48 hours. If it's only half full, the limit is 24 hours, so make a habit of packing empty space with cool-bag blocks and ice bags.

Food in the fridge will not last anywhere near as long as in the freezer. Beyond four hours, perishables are likely to be risky to eat and should be thrown away. The important thing with fridge perishables is how long they're above 4°C. More than two hours above that temperature and they are likely to be unsafe. Plan your next couple of meals based on the food that is going to be too warm soon. (This assumes you can eat it cold, or have some way of cooking it.)

If it looks as if the power is going to be out more than three or four hours, you will need to act to keep fridge food cool. Open the freezer for as short a time as you can and take out sufficient cool-bag blocks and ice bags, if necessary re-arranging the food to minimize air gaps; then close up the freezer again. Ideally, put chilled food into coolers or cool bags along with the ice and cool-bag blocks and, if there is room, put the coolers into the fridge. Pack insulation around the outside of the coolers – foam polystyrene, newspaper etc. – to minimize heat leakage.

When the power comes back on, check the temperature of your food. The frozen food should still have ice crystals in it and not be above 4°C. You will need a food thermometer to check this. If that's the case it's OK to refreeze (though the texture of some foods might suffer). Any perishable food from the fridge that has been over 4°C for more than two hours should be thrown away. If you were out during the blackout and don't know for certain how long perishables will have been above the safe limit, it's better to throw them away. See the safe food checklist on pages 81–6 for details of whether to keep or throw away specific foods.

POWER TIP
Emergency terracotta coolers

If your fridge is getting too warm, you can use evaporation to help keep perishable goods cool. When a liquid evaporates it takes heat from its surroundings – this is why sweat cools your skin.

There's a type of wine cooler that works similarly: an unglazed terracotta pot that is in water. As the water evaporates, the air inside the pot cools down, keeping the wine bottle inside cool. You can get a better and more long-lasting effect by taking two unglazed terracotta pots (plant pots, for example), one smaller than the other. Plug any drain holes in the pots so water doesn't leak out. Place the smaller pot inside the larger one and fill the gap between with coarse sand. Then pour water into the sand until it is saturated. You will get significant cooling as the water soaks through the terracotta and is evaporated. Put a piece of wet cloth over the top and items in the inner pot will soon be chilled. One of these dual-pot systems can run a couple of days between recharges of water. They work best out of the sun – full sun results in the water being lost too quickly.

Eating and drinking safely without power

Safe without cooking

One of the essential uses of power is to cook food, both to make it tastier and to make it safe to eat. With many foods, cooking is required to kill off the bacteria that cause food poisoning: this is true of most raw meats. In other cases, cooking makes the food more palatable and easier to digest. Then there are some foodstuffs for which cooking is essential to remove natural toxins – kidney beans, for example, are highly toxic unless cooked. (Tinned kidney beans have already been cooked.)

Although the usual restrictions apply in terms of making sure perishable foods are kept cool enough and eaten within their best-by date, a good range of foods can be eaten without cooking – and not just foodstuffs that are normally eaten cold anyway. Tinned baked beans are a classic case. Contact the manufacturer if uncertain.

Heating food

Although there is still plenty we can eat without heating it, being without power doesn't mean that it's impossible to heat up food – and if the weather is cold, eating hot food can be a great way to thaw yourself out. A small number of products are now produced in self-heating packages – most popularly coffee and soups. Otherwise you will need to apply heat externally.

You may have a camping stove or barbecue – bear in mind you should not use these indoors (see page 48). Alternatively, fires are the traditional way of cooking, either using an existing grate or stove in the house, or building a fire outside (see pages 46, 49). Use a pot suspended over the fire, grill over the flames, or bake by wrapping the food in foil and placing in the embers. You can also steam-bake, using multiple layers of well-soaked newspaper to wrap the food instead of foil – make sure the newspapers are kept damp to avoid them bursting into flames.

If it's a sunny day, it's perfectly possible to cook a meal without any fuel. There are many designs of solar oven that can be bought or made with scavenged materials. At its simplest, a solar cooker can be made from a pair of cardboard boxes, one larger than the other, around 40 cm square or bigger. Line both boxes with aluminium foil (shiny side out), then put one box inside the other with an air gap between, using balls of newspaper to support the inner box. Fix the boxes together using the flaps of the inner box and tape. You'll need a piece of card to go under your cooking container at the bottom of the box – this too should be foil-covered, but it's even better if you can paint this piece of foil with a non-toxic black paint. Cover the top of the oven with a transparent lid to keep the heat in but let the light through – heat-resistant glass or a roasting bag work well. Finally, make a second lid covered in foil, at an angle to catch the sunlight and reflect it into the box.

Remarkably, the temperature in the solar oven can get as high as 150°C at the peak of the heating cycle. Typically solar cooking will take around twice the time of a conventional oven – it's important to use a meat thermometer if cooking meat this way to ensure it's cooked through. See the appendix (page 257) for details of where to find instructions for building a whole range of solar ovens.

Pumped water?

If your water supply is pumped, you may lose it in a large-scale blackout. See page 94 and 98 for ways to find water and make it safe.

Communicating without power

Mobiles down

Landline telephone companies usually have their own separate power supplies, which feed through the phone network, so landline phones are less likely to be shut down by a blackout than mobile phones, which rely on power getting to your nearest phone mast. If your mobile isn't working, try an ordinary landline phone first. If your phone is cordless it will probably require mains power to work. Even if you don't normally use a corded phone, keep one in the cupboard for emergencies.

Landlines down

The chances are that if the landlines are down, the mobile network will be too, but it's worth trying your mobile if the landline is out. If the mobile network is working but not the landlines, you will be fine for speech, but denied your usual access to the internet via your computer. You can often use a mobile phone as your route into the internet, either directly or acting as a modem for your computer.

Many modern phones and computers, particularly laptops, use Bluetooth, a very short-range wireless network, to connect. If you have Bluetooth on your phone but not on your computer, you can buy a cheap add-on that plugs into a USB port on the computer and enables the two to talk to each other. Alternatively, many phones and laptops have an infra-red port, which can also be used to beam information between the phone and the computer and so connect to the internet.

To use your mobile as your internet connection you will need to get your computer to think of it as a modem. Check your mobile phone documentation. The Windows help system has information on using a Bluetooth phone to connect to the internet – search under 'Bluetooth mobile'. It makes a lot of sense to try this before you need to use it for real – it is much easier to use a mobile to connect to the internet once you have completed the initial set-up.

If you are without mobile and landline, try the local phone boxes. It's also worth trying to drive out of the affected area if you need to get in touch with someone urgently. You should then be able to use your mobile. Take a laptop with you, if you have one, for internet access. Again you may be able to use your mobile, or alternatively, if you have a wireless network connection on your laptop, find an affluent street with a lot of houses or offices and try to connect to a network. You will usually find that there are a large number of wireless networks on offer, and the chances are that some won't have security protection, so you can use them to access the internet.

If you have internet telephone software like Skype on your laptop, you can then also make use of this connection to make calls. Provided you use the network only to access the internet, while this is morally a little dubious, it isn't going to do any great harm to the network owner. Failing that, internet cafés aren't just for holidays – once you are out of the blackout area, they offer a good communications base.

This is fine for communication with someone outside the blackout zone, but won't help much with communications to anyone who is also blacked out. Consider investing in some walkie-talkies (also known as two-way radios). These are cheaper than they used to be, and have substantial ranges of between three and eight kilometres – excellent for local communications when the power is down. That apart, you will have to fall back on face to face and writing notes.

Protecting computers

Power cuts can result in your losing crucial information from your computer if it is shut down before you have saved data. You can also sustain damage to the computer itself and to the more delicate peripherals, particularly modems and broadband boxes, in the brief surge of power (called a spike) that often shoots through the grid when there is a lightning strike or when the power restarts.

To prevent damage, fit a surge protector. This looks like a multi-way power extension cable, but has built-in electronics to smooth out the impact of surges. Make sure your computer itself and other delicate electronics are plugged into a surge protector rather than a conventional socket. Surge protectors are not expensive, and well worth using.

If it's important to prevent losing the documents you are working on when the power goes, and you use a desktop computer, consider an uninterruptible power supply (UPS). These cost more than a surge protector, but are still not hugely expensive. The UPS contains a battery that will keep your essential equipment running for a few minutes so you have a chance to shut the computer down safely without losing anything. Most UPS units also have a piece of software that will automatically shut the PC down for you, though you don't have to use this.

A UPS also gives you the opportunity to send a few emergency emails and take a quick scan of the internet for information on what is happening before the power runs out. Make sure that as well as your computer itself, the screen and your modem/broadband box are plugged into the UPS. It's possible that the power cut will be so far-reaching that the telephone company loses power too, in which case you will lose your connection, but otherwise the UPS should give you a few minutes emergency internet access.

Staying safe in the home without power

Basic lighting

Lighting isn't just about getting things done outside daylight hours, it provides a sense of security. We are so used to being in our cocoon of power that when it is taken away it's easy to feel threatened, particularly at night. Lights help deal with this.

Torches are great for the first minutes of a power cut, or to get around where you don't have any other form of lighting, but they aren't very practical for more sustained use. Chemical glow sticks are even worse. They are fine to have as an emergency back-up to a torch to find your real lighting in the dark, or as a night-light, but they are useless for practical lighting. Consider getting some electric lanterns. Gas-powered camping lights should not be used inside the house because of potentially dangerous fumes. A good supply of batteries is essential to keep your electric lanterns going. Have some ordinary batteries for immediate use, but the combination of rechargeable batteries and a solar-powered battery charger will stand you in good stead for the long haul.

Even camping lanterns don't give the sort of light we are used to – a more powerful alternative is to combine a caravan-style car battery with low-energy bulbs, which will give you much brighter light for a long period of time.

POWER TIP
Lead acid batteries

The biggest problem with getting electrical power in a power cut is that it is difficult to store up electricity for future use. Battery technology has a long way to go before it's really up to the job. For the moment, the best option to get a sizeable amount of power is a lead acid battery, like a car battery.

1. **Two types** – There are two different types of lead acid battery. The type used in a car is designed to give a vast amount of power in very short bursts. They aren't intended to be run low, and will give up after a few recharges if they are run down a long way. Batteries like this are only for emergency use. For planned use, get a 'deep cycle' battery, such as a caravan battery, which is designed to be run down, then recharged, many times.

2. **Keep them topped up** – A lead acid battery gradually loses power over time. If you are keeping a battery as an emergency power source, use a trickle charger to keep it topped up – ideally a solar-powered charger.

3. **Consider full solar charging** – A trickle charger keeps a battery topped up when in store, but won't recharge it after a heavy evening's use. If you expect longer-term power outages, consider a solar charger that can fully recharge the battery.

4. **Use it occasionally** – Batteries that are never used will gradually build up a chemical layer on their plates that drastically reduces their efficiency.

5. **Stay safe** – In charging a lead acid battery, the highly inflammable gas hydrogen is produced. Make sure you charge the battery in a well-ventilated space, with no naked flames or sparks.

How much power? A typical caravan battery provides between 60 and 120 amp hours at 12 volts. If you are running an 11-watt low-energy light bulb (the equivalent of a 60-watt conventional bulb), you should be able to use it for about 60 hours off a 60-amp hour battery.

Living with candles

Most of us have candles around the house, and it would seem silly not to make use of this readily available light source in a blackout, but bear in mind that many fires were started by candles and oil lamps in the old days before electricity, and people then were used to living with naked flames – we are out of the habit, so need to be extra vigilant.

Candles don't give out a huge amount of light, so don't waste what they do produce. Put a mirror or piece of reflective foil behind the candles to reflect otherwise wasted light back into the room. As a bonus, if it's cold, candles will provide a small amount of heating.

⚡ POWER TIP

Candle drill

If your family ends up going to bed by candlelight it can be quite romantic – but don't let the romance conceal the increased risk of fire. Make sure any candles in children's bedrooms are positioned securely, can't be knocked over and are extinguished before you go to bed yourself. If your children need a night-light, don't use a candle, even a tea light – it's just too risky. Use a torch or a glow stick. The American Red Cross recommends never using candles in a blackout as the fire risk is too great. This seems a little excessive, but it is important to be aware of the considerable fire risk if you don't use caution.

Nature's lights

If you live in the countryside and don't have any candles there are a number of natural alternatives. Birch bark, with a natural oil content, burns well and can be made into a torch by rolling up sheets of bark, or placing a folded strip in the end of a stick. If you have some animal fat, you can render it down by heating it, then use it either to soak a natural taper like a cat's tail seed head, or to fill a lamp, which for these purposes is little more than a bowl with a wick made out of string or a rush stalk. These lamps are smelly and smoky, but do work.

Apply all the usual candle precautions to natural lights – more so, as they burn less predictably than a candle.

Transport issues from loss of power

Spotting fuel shortages early

Most of this section is concerned with loss of power to the home, but petrol and diesel shortages are also a way we can lose power. There are a number of circumstances when it's sensible to keep your fuel topped up before shortages hit. Watch out for:

❏ Reports of disruption in oil-producing countries

❏ News of difficulties with fuel supplies

❏ Warnings of electricity blackouts (remember petrol is pumped electrically)

In these circumstances, top up more often. Instead of letting your fuel tanks get down near empty before refilling, keep your car at least half full of petrol or diesel. Remember also, should you have a generator, to keep your fuel stocks for that topped up.

Dealing with long queues

Once fuel shortages bite there are likely to be long queues at petrol stations, wasting a huge amount of time and leading to frustration and anger. Bear in mind a few simple tips to minimize the impact of a fuel shortage:

❏ Don't go out to get fuel when the shortage has just been on the news – customers will rush to the pumps lemming-like and cause long delays.

❏ If at all possible, anticipate (see 'Spotting fuel shortages early' above).

☐ Use filling stations that are less popular. A few pence extra on the price is worth suffering to avoid a huge queue.

☐ Go at unusual times.

☐ If you are stuck in a long queue, don't keep your engine running.

☐ Make sure you have something to do. Take a talking book or something else to occupy your mind while waiting.

☐ Minimize your fuel use to avoid having to refill:

◆ Avoid sudden acceleration and braking. Smooth driving can reduce fuel consumption by up to 30 per cent.

◆ Try not to exceed 50–60 miles per hour – driving at 56 mph uses 25 per cent less fuel than 70 mph. Similarly, 70 mph uses around 30 per cent less than 85 mph.

◆ Take bike carriers and luggage racks off the roof to reduce drag.

◆ Empty the boot – don't carry around unnecessary weight. If you drive a car with removable seating, take out spare seats.

◆ Switch off the air-conditioning – this uses extra fuel, as do electrical extras like the heated rear window.

◆ Close your windows – open windows cause significant extra drag (in fact, open windows lose more fuel than air-conditioning uses).

◆ Avoid short journeys – these consume up to 60 per cent more fuel per mile than journeys where the car has a chance to warm up.

◆ Switch off the engine when in a traffic jam.

◆ Make sure your tyres are at the proper pressure. Under-inflated tyres waste fuel.

◆ Drive in the highest gear that's practical. A typical car at 37 mph uses 25 per cent more fuel in third gear than it does in fifth.

◆ Don't coast in neutral to save fuel in a modern, fuel-injection car: they use less fuel decelerating in gear than they do in neutral.

◆ Walk when you can.

⚡ POWER TIP
Alternative fuels

If petrol and diesel are in short supply, there are some alternatives, though you may find that using them isn't without problems.

The most common alternative fuels are LPG (liquid propane gas) and CNG (compressed natural gas). Several manufacturers offer dual-fuel vehicles that can handle one of these and petrol, or it's possible to have an existing car converted. Dual-fuel has the drawback of needing space in the car for two fuel tanks, but means you can still fill up with petrol when it's available – an advantage since there aren't as many locations supplying alternative fuels. This lack of filling stations also applies to biofuels like hydrogen, ethanol and methanol. Most current ethanol vehicles run on a petrol/ethanol mix, so still need some petrol to function.

Some diesel cars can be converted to run on vegetable oil. There are minor practical disadvantages to this – vegetable oil is thicker than diesel, so needs heating before it can be used in winter, and the smell is not to everyone's taste – plus you have to haul 50 litres of vegetable oil home from the supermarket; but this oil may be available when diesel is in short supply. (Only may be – your vegetable-oil supplier probably uses diesel trucks to carry its stocks.) Vegetable oil is also cheaper to buy than diesel, though countries with fuel taxes usually expect you to pay the duty.

Increasing use of alternative car fuels will eventually mean less dependence on oil, but there is a long way to go before this represents a practical solution for many people.

Producing power yourself

Sun and wind

One of the technologies recommended to reduce the impact of global warming – renewable energy sources – can also help those who are being hit by the consequences of climate change. Many of the devices for harvesting renewable energy – wind turbines, solar panels and water mills – are now available for domestic use. In many countries you can buy a wind turbine or solar panels from your local DIY store (though they need to be professionally fitted).

Solar panels come in two kinds: photovoltaic, which produce electricity, and hot-water panels that supplement your standard hot water supply. Although the water-heating variety is great to reduce your energy costs, they won't be any help in a blackout as they don't generate power and won't work without mains input.

Look into the options for micro-generation. They tend to be expensive (grants are available in some countries), but can offer a significant amount of power that is independent of other sources. Make sure you research this thoroughly, however. Watch out for some common problems:

❏ Some renewable power sources – many wind turbines and solar panels, for example – will only work if mains power is available, as they use a mains-powered device to manage the variable output of the generator. This means they are useless in a power cut. Generally speaking, if you have a generator that is set up to feed into the mains (sometimes called a 'grid-tied' system) it won't work in a blackout. Look instead for a stand-alone system that charges batteries.

❐ Check with your local water authorities if you have a stream running through your property and want to use a water turbine to generate electricity. Many authorities require these to be licensed or ban them altogether as they can have an impact on water flow further down the stream.

❐ Some local authorities require planning permission for erection of devices like wind turbines and solar panels. Check first.

❐ Wind turbines can be noisy. Make sure you have a site that won't cause noise pollution – and a location that catches the wind. This may seem obvious, but even full-scale commercial wind-powered generators have sometimes been sited where there isn't enough wind to produce worthwhile results.

❐ Wind turbines that are small enough to be fitted to your house will produce only a fraction of your requirements and are unlikely to pay for themselves: consider them only as an emergency power source, not as a money-saving alternative to buying electricity from the grid.

Human power

One way of looking at people is that they are mechanisms for turning chemical energy stored in food into the mechanical energy we use, amongst other things, to walk and move. It's entirely possible to go one step further and turn that mechanical energy into electrical energy by using a human-powered generator. Small examples of these are available as 'wind-up' radios and torches. The great thing about these is that you aren't dependent on batteries which can run out.

When you turn the handle on a wind-up torch you are running a very small manual generator. These can be scaled up to provide more power, typically run from a bicycle-style mechanism. Instead of getting yourself a cross trainer or exercise bike, consider getting a pedal-powered generator. You can use it to charge up large batteries that can then be used for lighting and other relatively low-power appliances. A pedal-powered generator can output around 60–70 watts – a one-hour workout would enable you to run a 10-watt low-energy light bulb for a whole winter evening. At the time of writing there wasn't much available commercially in the way of pedal-powered generators – most users of pedal generators have rigged them up themselves – but it's only a matter of time before the makers of exercise machinery wake up to the possibilities.

Inverters

Storing electricity is a messy business, hence the lack of any really satisfactory electric cars. About the best we can do at the moment is the lead acid battery (see power tip on page 58). If you have one or more charged-up lead acid batteries, ideally the 'deep cycle' batteries often called caravan batteries or leisure batteries, then you have a potential source of power that will last you at least an evening. The only problem is getting the power from a 12-volt battery into the appliance you want to use, like a low-energy light.

What's needed is an inverter. This is a little device that takes the 12-volt DC output from the battery and turns it into the 110 or 240-volt AC you need to run your household gadgets. Inverters come in different ratings, from the very cheap ones at around 75 watts – plenty to run a laptop or a low-energy light bulb – up to around 500 watts, which still aren't particularly expensive. Bear in mind, though, that a typical battery will only produce this power level for an hour or two – it's best to stick to low power uses.

If you haven't got a separate deep cycle battery, you can buy little inverters that plug into a car's cigarette lighter or auxiliary socket, but this should be a last resort as you will significantly shorten a car battery's life if you use it this way for long periods of time.

Generators

If you are lucky enough to have a generator, you will be able to continue to run some electrical devices as long as you have fuel, but remember that generators are limited in their power output. Identify your key requirements – they might be:

❏ Power to keep gas/oil central heating running

❏ Power for a fridge or freezer

❏ One light

❏ Some use of TV or radio and computer

❏ Air-conditioning unit

❏ Heating food and water with a (relatively) low-energy electric ring

Exactly what you can do depends on the capabilities of your generator. Check the power output and add up the watts used by the different appliances. Electric cookers, kettles and heaters all take a very large amount of power. An electric kettle, for instance, can take as much as 2,000 watts, while cookers and hobs will take a lot more. You can get lower-rated kettles and small electric rings – check the packaging for power consumption – but you may well be better off using gas for cooking. If you are using a high-rated appliance, make sure there is nothing else running at the same time.

The essential number to check is the power rating, measured in watts. A generator might be rated at 2 kW (2,000 watts). Some appliances might tell you what they need in amps. To find watts, multiply amps by voltage. The mains voltage is 220 to 240 in Europe, 100 to 120 in Canada, the USA and Japan. If you have an appliance that uses 5 amps in Europe, that's around 1,200 watts; in the lower voltage countries it would be around 550 watts. Be particularly careful with the power rating of your microwave. The microwave will take more power from the mains than its cooking rating. A 900-watt microwave will actually consume 1,200–1,400 watts in electrical power.

Unless your generator has been professionally wired into your mains with an appropriate control box, don't try to use your mains wiring to get the power from your generator to the appliances. If you want to use your mains, get an electrician to fit a power transfer switch. This enables you to choose whether power comes from the electricity grid or your generator – otherwise, your generator will try to feed power back into the grid. Not only is this wasteful, it can be highly dangerous for anyone attempting to work on the mains elsewhere. If possible connect appliances directly into the generator, otherwise use a suitably rated extension cable.

POWER TIP
Extension overheat

Extension cables can carry much less power when still on the reel. The reel of wire acts as a coil and will heat itself up. A typical 13-amp extension cable when reeled may take only as little as 5 amps (most reels are marked with the maximum rating when fully wound as well as unwound). If you run a high-power device for any length of time, the cable will get so hot it starts to melt and the resulting short circuit could cause a fire. For safety, always unwind the reel completely, even if you don't need all of the cable. Make sure the cable is tucked out of the way to avoid tripping.

Generators themselves present a number of safety hazards. They will usually be petrol- or diesel-driven, which involves a risk of fire. Like a car exhaust, a generator's exhaust fumes contain poisonous carbon monoxide. Never run a generator in an enclosed space, even if it is well ventilated. Don't use them in the house, porch, garage or basement. Carbon monoxide is an invisible, odourless killer. You can't risk exposure to it.

Unfortunately, it's not a good idea to get a generator wet, either, so provide it with some kind of open canopy to keep it dry. Make sure you are dry yourself when you touch it. A generator is producing mains-rated electricity, and this can kill.

Resources

Cars

If you own a car, it will provide significant resources in the event of a medium-term power outage. At the very least it has a powerful battery, and acts as a petrol-driven generator. If it is very cold and all other heating has failed, you can heat the car and use it as a survival environment. Bear in mind that the car heater does not provide heat directly, it only channels heat from the engine, so the engine has to be run for a minimum of 15 minutes on a regular basis to keep up a supply of heat. This will also help recharge the battery.

If you are using your car heater to stay warm, switch it to recirculated air (otherwise you are constantly drawing in cold air), and direct the air to the footwells – if the air is directed on to the windscreen, a lot of heat will be lost through the glass. Use available insulation (see 'Foam polystyrene' and 'Loft insulation' opposite) to reduce heat loss, particularly from the windows.

You will probably have a radio – helpful both for entertainment and to hear any emergency messages. Your car's power can also be used to provide lighting, to heat drinks, and to run other electronic devices. Bear in mind that the battery must be recharged on a regular basis, as you need enough power to get the engine started, and car batteries do not like to run low on charge.

 Never run a car engine while in a confined space (a garage, for example). The build-up of exhaust fumes can kill.

Caravans

Many drivers despise caravans because they hate being behind them on the road, but caravans are usually well equipped for surviving blackouts because they are set up to operate independently from mains supplies. In a caravan you may well have bottled gas and batteries to provide power, allowing you to continue living in relative luxury for days during a blackout. That might mean light, heat, hot water, cooking, fridge, TV or radio – all without a mains connection. Remember not to let your gas supply run low, and keep your battery or batteries topped up with a solar trickle charger.

Foam polystyrene (styrofoam)

Foam polystyrene, widely used as a filler in packaging (in the form of thumb-sized pellets, flat sheets, or pre-moulded shapes – usually white, always very light), is a very good insulator. If you store old delivery boxes and similar containers in your loft or garage, you may have a significant amount of foam polystyrene available to help reduce heat loss from windows, to keep food cool in the fridge or freezer, and to keep hot items warm for longer.

Heating systems with gas/oil tanks

 This suggestion should only be carried out by a qualified engineer, or in the event of total breakdown of society.

If your heating is provided by a liquid gas (e.g. propane) or oil tank, you have a self-contained energy store, independent of mains supplies. In the event of a long-term loss of power, this could provide power for cooking and lighting (gas only) as well as heating. This would require modifications to the system, both to enable a limited part of the heating to run without electricity and to enable a cooking ring and light to run off the supply, but in a long-term emergency this should not be overlooked.

Loft insulation

 The material in this resource can cause irritation. Use gloves and a face mask when handling.

Many houses have insulation in the roof space to avoid heat loss. If you have urgent need for insulation to keep something cold or warm, loft insulation is designed for the job. Don't allow the insulating material (usually fine glass fibres, which penetrate the skin and cause irritation) to come into direct contact with skin, wear a mask to avoid inhalation, and don't let it come into contact with food; make sure it is securely contained in a bag, such as a dustbin liner.

If power outages last for a considerable time you may wish to conserve what resources you have and heat only one room. If this is the case, look at opportunities for moving loft insulation to surround and insulate that room as much as possible (particularly windows, but also walls, ceiling and floor).

House and garden

Make a sweep around your house and garden looking for appropriate resources. Look for safe flammable materials, such as wood, and insulating materials to keep yourself warm or to keep chilled food cool. Establish just what you have available to eat (and, in the event of losing pumped water, to drink). You may also need items like duct tape or packing tape to fix insulation in place. By looking at your house and garden (not forgetting any store cupboards, the garage, loft and attic if you have them) as a source of resources rather than a home you will spot items you might otherwise miss.

Swimming pools

If you have access to a heated swimming pool, particularly an indoor pool, this will act as a heat sink during the power cut. A well-insulated indoor pool at around 28–29°C will lose only about a degree a day. It's not exactly hot bath temperature, but can provide useful warm water for washing and to warm up chilled extremities when you have been out on a cold day. Do not drink swimming-pool water without appropriate treatment (see page 94).

CHECKLISTS

Scavenging

Ideally you should have prepared a kit for coping with power cuts and energy loss (see below), but it may be you haven't got round to it, or aren't at home. The scavenging checklist has suggestions for finding essentials around you.

Check through the resources section (pages 70–72) for ideas .... ❏

Collect insulation, especially if it is cold .... ❏
You will need this for yourself, your hot-water tank, your chilled and frozen food, and your warm room. As well as foam polystyrene and loft insulation (see page 71), collect blankets, duvets, old coats, newspapers – anything that can be used to keep heat in or out.

Find objects that make a light .... ❏
Look for battery-operated electrical items with bright lights like mobile phones and PDAs, plus flame-based lights like matches and candles.

Search through your garage, loft and basement .... ❏
There may be forgotten items – camping stoves, old torches and candles, for example – that you can use to make light and keep warm.

Basic kit

You can put together a simple power-cut and energy-loss kit, to make it easier to cope when you lose power, without spending a fortune. Make sure the most essential items are stored somewhere easily accessible, even in the dark. Try feeling your way to the storage location with your eyes closed.

Torches (battery and at least one wind-up) .... ❑
Have these placed strategically around the house. Make sure at least one is wind-up so you aren't caught out without batteries. Chemical light sticks can also be used as emergency lights to spread around the house to help you get to your torch.

Candles and matches .... ❑

Waterproof (lifeboat) matches and cigarette lighter .. ❑
If you have to make a fire out of doors (see page 49), then it's essential to have matches that aren't overwhelmed by wind or damp weather. A pack of waterproof matches, sometimes called lifeboat matches, will stand you in good stead.

Warm clothing and covers/sleeping bag .... ❑

Battery radio .... ❑

Electric camping lantern .... ❑

Camping stove and gas .... ❑
For external use only: avoid using in the house.

Cool-bag blocks .... ❑
The plastic blocks used to keep coolers and cool bags cold will also help your fridge or freezer keep cool longer. Routinely keep empty spaces in your freezer crammed full with these (freeze bags of water if you haven't enough). A full freezer will keep food twice as long as a half-empty one.

Insulating material to keep frozen food cold .... ❑

Cool bags or coolers for chilled food .... ❑

Reasonable supply of canned and dry food and milk ❑

Food thermometer (preferably digital, quick response) .... ❑
Perishable food from the fridge that goes over 4°C may only be safe for around two hours. Check the temperature, ideally in the middle of the food, and check the appearance and smell of any perishables.

Phone that works without power .... ❑
Landline phone lines often still work in a power cut as the phone system has a separate power source. But many modern phones (cordless, for instance) only work if connected to the mains. Make sure you have at least one phone that functions with no mains power.

Mobile phone .... ❑
Major power cuts may also take out mobile phone systems, but it is worth having one accessible in case it still works, and the backlight will act as a torch.

Spare batteries .... ❑

If you want to go a little further

Wind-up radio .... ❑
The chances are your battery radio will be OK, but a clockwork radio will ensure you can keep up to date with what's happening even if you forgot to replace the batteries in your other radio.

Uninterruptible Power Supply (UPS) for computer .... ❑
When there is a power cut, whatever you were working on since your last save will be lost, and other files may be corrupted. A UPS is a special extension lead with built-in batteries that will keep your computer running for a few minutes into a power cut so you can shut down safely.

Deep cycle 12V battery (caravan battery) .... ❑
Like a car battery, but designed for long-term, low-drain use, rather than the quick and heavy use required of a car battery.

Car battery solar trickle charger .... ❏
Keeps the battery topped up when you aren't using it.

Walkie-talkies (two-way radio) .... ❏
Great for local communication when there's no power.

Inverter .... ❏
To power mains devices from a caravan battery.

**Rechargeable small batteries and solar
battery charger** .... ❏
*You can now buy reasonably priced chargers for rechargeable
batteries that run off solar power, keeping your small powered
devices going in the event of a long-term power outage.*

Advanced kit

If you want to take this seriously, you can go even further.
Some of the options here may cost as much as a small family
car – but the long-term benefits are considerable. The ideas
in the advanced kit will provide benefit even before serious
climate-change impact. A generator, for instance, will come in
useful if you suffer from power cuts, a renewable energy
source can be used to reduce your electricity bills, and a
stove makes a great feature for the house.

Multi-fuel stove .... ❏
*If it's practical in your house, have a multi-fuel stove fitted. This
can burn wood and coal/smokeless fuel. Such a stove is safer
and more efficient than an open fire and can be used with
scavenged wood if no conventional fuel is available.*

Portable heater(s) .... ❏
*If a stove isn't practical, or to heat other parts of your house,
make sure that you have one or more indoor portable heaters
(see page 47) to provide instant heating on demand.*

Well-insulated hot-water tank .... ❑

If your tank isn't already well insulated, get it lagged or replace it with a modern insulating tank, so you will keep your hot water longer in the event of a power cut.

Home insulation survey and upgrade .... ❑

Get your house checked for weaknesses in insulation (windows, cavity walls and roofs, for example), and upgrade to reduce heat losses.

Generator .... ❑

A generator will ensure you can provide power for essentials, such as operating your gas heating system and keeping your fridge and freezer running.

Renewable energy source .... ❑

Solar panels, wind turbines and water-powered generators may give you the ability to generate your own power without even needing to keep a stock of petrol.

Car-battery full solar charger .... ❑

If you don't want to run to a full-scale renewable energy source, at least consider a solar car-battery charger that can restore the charge, rather than just a trickle charger – a small-scale solar-panel solution.

Caravan .... ❑

As we've seen, a caravan provides a living environment that works isolated from mains power. Consider this as a safety net.

After blackout warning

Not all power cuts come out of the blue. In the event of fuel shortages there may be 'rolling blackouts' where towns or sections of a city have planned blackouts for a period of time. This checklist itemizes some essential preparations for planned blackouts.

Fill your car up with fuel .... ❏
*Garages use electricity to pump fuel, so you won't be able to fill up
in a blackout. Stock up for any generator you might have as well.*

Check food and water supplies .... ❏
*Without power you may find it harder to get to the shops, many
of which may not be open.*

Make sure your freezer is well packed .... ❏
*If you haven't enough food to fill it, pack it out with cool-bag
blocks and bags of ice to make it longer lasting in a blackout.*

Keep a supply of cash .... ❏
*ATMs won't be functioning and banks usually close in a blackout.
Have some cash ready for emergency requirements.*

Carry a small emergency kit .... ❏
*At the very least make sure you have a torch with you at all times,
and keep a torch, blanket and first-aid kit in the car.*

Make sure you have the basic kit (page 73) .... ❏

Consider items from the advanced kit (page 76) .... ❏
*If blackouts are going to be more common, think about the
benefits of these larger investments in self-sufficient power.*

Environment check

If you are in a strange place when there is a power cut, it can
be extremely disorienting. Knowing how to react can make
the difference between life and death if people begin to
panic. Use these short mental checklists to be prepared.

In a windowless space (cinema, theatre etc.) the big problem
is going to be finding your way out. There should be
emergency lighting that isn't mains-dependent, but you will
still be disoriented.

As a habit, when you get in your seat, count the
number of steps to exit level (if on raked seating) .... ❏

Memorize two routes to emergency exits .... ❏
If a blackout does occur, wait for the initial panic to subside.
Feel your way to the end of the row of seats, then count down
steps to the exit level. Try to get close to the wall to feel your
way along towards the exit. You are likely to be jostled by
other people. Keep your centre of gravity low and walk as if
on ice to reduce your chances of being knocked over. Don't
run: running makes it particularly easy for you to be knocked
over and trampled.

In a normal space (workplace, restaurant, shop etc.) you may
have natural lighting to help. The essential is to have two
clear exit routes in mind – one to the way you came in,
another to an emergency exit. You don't have to go out of
your way to find this usually, just keep an eye out for green
exit signs and remember how you have moved around the
building. This is particularly important if it's a warren-like
building.

Memorize two routes to emergency exits .... ❏
In a complex environment like a hotel, especially one where
you are likely to be in an unprepared state such as sleeping,
on reaching your room spend a couple of minutes doing an
exit survey *before doing anything else*. This will be valuable
in case of fire as well as in a power blackout.

Find the nearest emergency exit and
check it isn't blocked .... ❏

Count doorways between the exit and your room ..... ❏

Ideally do the same for an emergency exit in
the other direction .... ❏

Leave your room key and a set of clothes somewhere easily accessible from the bed, and make sure you can get out of the room in complete darkness, in case there is a blackout while you are asleep .... ❏

⚡ POWER TIP

Memory techniques

Will you remember how many doors you counted to the emergency exit, or even which direction to go in the pitch dark, or in thick smoke? Use simple memory techniques to fix this information in your memory.

❏ Walk through the journey to the emergency exit in your mind, imagining you are touching each door as you go.

❏ Think of a similar route you are very familiar with (for example from your bedroom to the bathroom at home), with the same turn out of the door – use this to remember the direction.

❏ Learn these simple number rhymes, and use them to fix the number of doors in your memory by building a short but dramatic, colourful and exaggerated story in your mind:

One – **GUN**	Six – **STICKS**
Two – **SHOE**	Seven – **HEAVEN**
Three – **TREE**	Eight – **WEIGHT**
Four – **DOOR**	Nine – **LINE**
Five – **HIVE**	Ten (o) – **HEN**

For example, if you turn left and then count thirteen doors to the exit, picture yourself coming out of the door and turning left to find that you are facing down the barrel of a huge gun. It goes 'bang' and, instead of a bullet, a purple tree sprouts out of the end of the barrel. So that's one (gun), three (tree):

thirteen doors. Repeat this little story to yourself several times to fix the number in your memory. (You need to learn the rhymes first, but this doesn't take long and will come in useful for any other numbers you need to commit to memory.)

Safe food

FROZEN FOODS

Meat and mixed dishes:	Still contains ice crystals. Not above 4°C	Thawed, held above 4°C for over 2 hours
Beef, veal, lamb, pork, poultry, minced meat and poultry	Refreeze	Discard
Casseroles with meat, pasta, rice, egg or cheese base, stews, soups, convenience foods, pizza	Refreeze	Discard
Fish, shellfish, breaded seafood products	Refreeze	Discard
Dairy/eggs/cheese:	**Still contains ice crystals. Not above 4°C**	**Thawed, held above 4°C for over 2 hours**
Milk, cream	Refreeze	Discard
Eggs (out of shell), egg products	Refreeze	Discard
Ice cream, frozen yogurt	Discard	Discard

Soft and semi-soft cheese, cream cheese, ricotta	Refreeze	Discard
Hard cheese (e.g. cheddar, Gruyère, Parmesan)	Refreeze	Refreeze
Fruit and vegetables:	**Still contains ice crystals. Not above 4°C**	**Thawed, held above 4°C for over 2 hours**
Fruit juices	Refreeze	Refreeze. Discard if mould, yeasty smell or sliminess develops.
Home or commercially packaged fruit	Refreeze	Refreeze. Discard if mould, yeasty smell or sliminess develops.
Vegetable juices	Refreeze	Discard if above 10°C for over 8 hours.
Home or commercially packaged or blanched vegetables	Refreeze	Discard if above 10°C for over 8 hours
Baked goods and baking ingredients:	**Still contains ice crystals. Not above 4°C**	**Thawed, held above 4°C for over 2 hours**
Flour, cornmeal, nuts	Refreeze	Refreeze

Pie crusts, breads, rolls, muffins, cakes (no custard fillings)	Refreeze	Discard if above 10°C for over 8 hours
Cakes, pies, pastries with custard or cheese filling, cheesecake	Refreeze	Discard
Commercial and home-made bread dough	Refreeze	Refreeze

CHILLED FOODS

Dairy/eggs/cheese:	Food still cold, held at 4°C or above for under 2 hours	Held above 4°C for over 2 hours
Milk, cream, sour cream, buttermilk, evaporated milk, yogurt	Keep	Discard
Butter, margarine	Keep	Keep
Baby formula, opened	Keep	Discard
Eggs, egg dishes, custards, puddings	Keep	Discard
Hard and processed cheeses	Keep	Keep
Soft cheeses, cottage cheese	Keep	Discard

Fruit and vegetables:	Food still cold, held at 4°C or above for under 2 hours	Held above 4°C for over 2 hours
Fruit juices, opened; canned fruits, opened; fresh fruits	Keep	Keep
Vegetables, cooked; vegetable juice, opened	Keep	Discard after 6 hours
Baked potatoes	Keep	Discard
Fresh mushrooms, herbs, spices	Keep	Keep
Garlic, chopped in oil or butter	Keep	Discard
Meat, poultry, seafood:	Food still cold, held at 4°C or above for under 2 hours	Held above 4°C for over 2 hours
Fresh or leftover meat, poultry, fish, or seafood	Keep	Discard
Lunchmeats, hot dogs, bacon, sausage, dried beef	Keep	Discard
Canned meats NOT labelled 'Keep refrigerated' but refrigerated after opening	Keep	Discard
Canned hams labelled 'Keep refrigerated'	Keep	Discard

Mixed dishes, side dishes:	Food still cold, held at 4°C or above for under 2 hours	Held above 4°C for over 2 hours
Casseroles, soups, stews, pizza with meat	Keep	Discard
Meat, tuna, prawns, chicken, or egg salad	Keep	Discard
Cooked pasta, pasta salads with mayonnaise or vinegar base	Keep	Discard
Gravy, stuffing	Keep	Discard
Pies, breads:	**Food still cold, held at 4°C or above for under 2 hours**	**Held above 4°C for over 2 hours**
Cream- or cheese-filled pastries and pies	Keep	Discard
Fruit pies	Keep	Keep
Breads, rolls, cakes, muffins, scones	Keep	Keep
Refrigerator biscuits, rolls, cookie dough	Keep	Discard

Sauces, spreads, jams:	Food still cold, held at 4°C or above for under 2 hours	Held above 4°C for over 2 hours
Mayonnaise, tartare sauce, horseradish	Keep	Discard
Opened salad dressing, jam, relish, taco and barbecue sauce, mustard, ketchup, olives	Keep	Keep

Prepared by Giant Food, LLC., Landover, Maryland. Used with permission.

THREE

THE STAFF OF LIFE

Under normal conditions we can last without power indefinitely, however uncomfortable and unhappy it might make us. Going without food and water is a different ball game. Very soon, if supplies run short, this is a matter of life or death.

What to do ...

... if your water is cut off

... when food supplies run short

... to live off the land

... to get access to food

... to have a sensible diet

Food and drink on tap

It has become a natural assumption that we can turn on the tap and get water, or call in at the supermarket to stock up on whatever we need, whenever we need it. But water is becoming scarcer both for drinking and growing food. When chaos strikes, there is no reason to assume our water supplies will continue as normal, or that food will be so readily available.

We are all familiar with water shortages somewhere in the world – yet at first glance, the whole concept of running low on water is an insane one. Looked at from space, the defining feature of the Earth when compared with the other planets in our solar system is water. Our world is blue with the stuff. In round figures there are 1.4 billion cubic kilometres of water on the Earth. This is such a huge amount, it's difficult to get your head around. A single cubic kilometre (think of it, a cube of water, each side a kilometre long) is 1,000,000,000,000 litres of water.

Divide the amount of water in the world by the number of people and we end up with 0.2 cubic kilometres of water each. More precisely, 212,100,000,000 litres for everyone. If you stack that up in litre containers, the pile would be around 10 million kilometres high. With a reasonable consumption of 5 litres per person per day, the water in the world would last for 116,219,178 years. And that assumes that we totally use up the water. In practice, much of the water we 'consume' soon becomes available again for future use. So where's the water shortage?

Things are, of course, more complicated than this simplified picture suggests. In practice, we don't just get through our five litres a day. The typical western consumer uses between 5,000 and 10,000 litres of water a day.

The typical western consumer uses between 5,000 and 10,000 litres of water a day

In part this happens directly. Some is used in taking a bath, watering the lawn, flushing the toilet – but by far the biggest part of our consumption, vastly outweighing personal use, is the water taken up by manufacturing the goods and food that we consume. Just producing the meat for one hamburger can use 3,000 litres, while a 200g jar of coffee will eat up 4,000 litres in its production.

However, even at 10,000 litres per person a day, we still should have enough to last us over 57,000 years without even adding back in reusable water. So where is the crisis coming from? Although there is plenty of water, most of it is not so easy to access. Some is locked up in ice or underground, but by far the greatest part – around 97 per cent of the water on the planet – is in the oceans. It's not particularly difficult to get to, certainly for any country with a coastline, but it is costly to make use of. The fact that many nations with coastlines are prepared to spend huge sums of money on reservoirs to collect a relatively tiny amount of fresh water, rather than use the vast quantities of seawater that are available, emphasizes just how expensive is the desalination required to turn seawater into drinkable fresh water.

Water shortages, then, come down to a lack of cheap power – as, more indirectly, do limits on food. Drought makes food harder to grow, relying more on expensive irrigation, but with sufficient power, irrigation should not be an issue. On the world scale, as climate change bites, limits on power availability make it harder to provide irrigation and to transport food around the world to meet global need. Regular blackouts, as described in the previous chapter, will mean that pumped water can suddenly become unavailable. In a power cut, supermarkets can't operate their tills, freezers and chiller cabinets. The food distribution network relies on availability of diesel fuel, and of electricity to pump that fuel. Without power, food distribution grinds to a halt.

The limits on power have an indirect impact. The most direct consequence of climate change is a reduction in the local availability of food and water. Drought conditions due to global warming will be more common. Countries that have in the past got away with a hosepipe ban in the summer are likely to face more severe rationing of water. Countries that have historically had water shortages are moving towards crisis. At the same time, crops will tend to move away from the equator, so plants that, for instance, used to be common only in southern Europe or in the southern states of the US will thrive in more northerly and extreme southerly regions. Areas that already had difficulties sustaining food growth will find it even harder, while countries that rely on imported food could have problems even if they themselves aren't losing food crops.

By the time the food and water supply is at risk, it's probably too late to stock up in a big way. Even with a warning that troubles are coming, at some point you may have to rely on your own capabilities to find food and water. It's not enough to stockpile. Whether you are based in a town or the countryside, you need to know how to find what you need to survive and how to avoid being killed by eating and drinking the wrong things. One of the side effects of our easy-access, food-on-tap life is that we assume water will be safe to drink and food safe to eat. The real world is very different.

Solutions

Your water supply

A human being can go several weeks without food, but missing out on water will lead to death in just three or four days. (This is temperature-dependent. You can last several days longer by taking the water-use reduction advice on page 97.) Unfortunately, contaminated water has always been one of the most common sources of poisoning and disease. We tend to take safe water for granted – things may be different in the future. It's obvious that a murky garden pond is contaminated, but don't assume a clear, fresh-looking mountain stream is any cleaner. You can't see bacteria and floating parasites – it's always best to assume that any water source away from recognized drinking water is contaminated. Treat it, rather than risk infection.

⚡ POWER TIP
Water loss

Just sitting around in the shade at a relatively cool time of year you will lose around a litre of water a day. Doing normal activity you will be shedding two to three litres – and at high temperatures, or if you are ill, you may lose significantly more. All this has to be replaced.

Store and survive

If it looks likely that the water will be cut off, or it is in increasingly short supply, make sure that you have a good store of drinking water. Use your bath, kitchen pans and any large bottles and containers for storage. Open containers should be covered (find a clean plastic sheet or something similar to go over the bath). The cover is for three purposes – to reduce evaporation, to keep dirt and bugs away, and to keep out light to prevent algae growing. If you use clear plastic, put something lightproof over the top. If possible use water from the mains, rather than from a cold-water tank in the house, which may normally be used to fill the bath. If you have to use water from the tank, always filter and boil before use (see page 95). Even drinking water from the mains should be boiled after it has been kept for more than a couple of days.

If you have a chance, you can also dig yourself an extra water storage pit. Dig a large hole in a shaded area of ground and line it with a pond liner, plastic sheeting or cement (this needs to be dried and cured before use). Make sure it is kept covered, and if the cover is wood or similar material, fix a plastic sheet across its underside, to capture evaporated water, some of which will drip back down into your emergency supply as condensation.

Think before you drink: assume all water from an uncertain source is unfit to drink without being treated. Never drink seawater or urine – these both contain chemicals that will dehydrate you rather than replace your water levels. Both urine and seawater can be made safe to drink using a still or a solar still (see page 95). Even properly treated tap water won't stay safely drinkable for ever. In a sealed bottle, stored in a dark cool place, it will be safe to drink for about a month, but if not properly stored will need treating after a couple of days.

Meet the pests

It's worth familiarizing yourself with the potential contaminants that make your water undrinkable without treatment,
to be aware what you need protecting from. The vast
majority of these harmful additions to pure water are invisible.

❏ Bacteria – The most familiar infectious 'bugs' or 'germs',
tiny single-celled organisms that exist in vast quantities all
over the Earth. Many are harmless (we rely on 'friendly'
bacteria in our gut to help process our food), but some can
cause anything from a mild upset stomach to death. Among
the many bacteria that may be found in water are cholera,
salmonella (with varieties responsible for various kinds of
food poisoning and typhoid), *E. coli* and campylobacter.
Some very fine filters can remove bacteria, but the only sure
way of getting rid of them is boiling, distillation or appropriate chemical treatment.

❏ Viruses – We are most familiar with airborne viruses like
the common cold and flu, but these tiny infecting particles
can also be found in water. Viruses are technically not alive,
but use parts of the invaded body's cells to help them
operate. Polio, hepatitis and Norwalk virus are among the
waterborne infections that it's possible to pick up. Boiling,
distillation and some chemical treatments will deal with
viruses.

❏ Protozoa – Much less well known than bacteria and
viruses, protozoa are single-celled like bacteria, but significantly larger and can have quite complex features, like
biological propellers, to move themselves through the water.
Many protozoa are harmless, but a number produce a
dangerous infection of the gut and cause serious illness.
Best-known examples are cryptosporidium and giardia. If
protozoa find themselves in a hostile environment they
change into a secondary form called a cyst, which is very

resistant to attack and will survive most chemical treatments. Only boiling and distillation can ensure that protozoa are killed off.

❏ Parasites – This label covers a whole range of creatures that depend on another species as a host. Some parasites exist in a mutually beneficial state, like those friendly bacteria in the stomach, but most take from the host without giving back. Perhaps most familiar are fleas, lice and worms, but parasites can range from the microscopic creatures responsible for malaria and trichinosis all the way up to the huge tapeworms that live in the intestines and can be up to several feet long. Boiling and distillation will kill all parasites, and they are also usually removed by filtration.

❏ Chemicals – Water itself is a chemical, and technically anything else in it, non-living, is a chemical contaminant. This can be as simple as common salt, all the way up to complex organic chemicals that are water-soluble. Some chemical contaminants – sand, for example – don't dissolve in the water, but may be suspended in water, making it dirty. This is the one kind of contamination that won't be helped by boiling. Basic particles that aren't dissolved will be removed by filters, while distillation will take out all the chemical contaminants that don't have a lower or similar boiling point to that of water. The remaining contaminants are particularly difficult to get rid of. They can be removed by more sophisticated 'partial distillation' apparatus, that only lets through items with a particular boiling point, or by special filters – active carbon filters will remove a fair number of the organic chemicals that have lower boiling points than water, and even more can be dealt with using a reverse osmosis purification system.

Safe to drink

Ideally you will restrict your drinking water to known safe sources, but if the water supply is disrupted for any length of time, you will have to look for ways to make other sources of water drinkable. There are four ways to treat water: distillation, boiling, filtering and with chemicals.

Filtration will remove any solid pollutants in the water and will reduce some dissolved contamination (lead, for example). Make sure the water filter you are using is fresh. Filtering the water is not sufficient to remove bacteria and viruses, but will result in clearer, more palatable water which can then be boiled. As an alternative you can buy a water purifier which combines a filter with a chemical agent to kill bacteria: this is particularly helpful if you need to use it away from the house where you may not be able to boil the water. There are sophisticated filters that are designed to remove the volatile chemicals that even distillation can't get out of the water – these usually combine active carbon and reverse osmosis.

The sort of water filter used in hard-water areas to stop kettles furring up is often a carbon cartridge, but is not designed to make contaminated water drinkable. Check manufacturers' details, but filter effectiveness normally goes up from filter paper, through carbon cartridges, ceramic cartridges and finally a combination of activated carbon and reverse osmosis. If you don't have anything else to filter with, a tightly woven piece of cloth formed into a bag will get out the larger bits and pieces – though you should always boil the water as well.

Boiling is an effective and simple way to kill off bacteria and viruses. It is important to ensure that the water boils properly – for one minute if clear and ten minutes if cloudy. Do not rely on an electric kettle which brings water to the boil then switches off; this won't boil the water long enough; use a conventional kettle or a pan with a lid. You will then need to leave significant time for the water to cool (it should be covered during cooling). The effectiveness of boiling drops with high altitude (where water boils at a lower temperature). Increase boiling time by about a minute for each 1,000 metres above sea level.

An alternative technique to purify water is distillation.

This involves boiling the water, letting it escape as a gas (water vapour), then cooling down the gas so it turns back to water. Any solids, bacteria and viruses will be removed by distillation, so it is a very effective method. It will also remove many contaminating chemicals that boiling and filtering might miss, making it an essential process if you have to make usable water from (for example) swimming pools which may have had significant amounts of chemicals added. The only hazard that gets through untouched is other substances that are liquid at room temperature, such as dry-cleaning fluid, which will boil along with the water then be condensed alongside it. There are distillation mechanisms to deal with these as well, but you would need a commercial still to cope with these.

For basic distillation, start by boiling the water normally for a couple of minutes to let any volatile contaminants boil off. Then cover the pan, collecting the vapour through a tube in the cover, running that tube through something cold to a lower collection vessel. You can also use a solar-powered still – see 'Water from the Sun' (page 99).

Chemical treatment can also be used to kill bacteria and viruses (look for water purification tablets or kits), but is second best to boiling or distillation. If it is absolutely impossible to boil water and you have no purification kit, 2 drops of household bleach (simple bleach, not cleaners containing bleach) or 3 drops of 2 per cent tincture of iodine per litre of water will help reduce the bacterial and viral content, but this should only be used as a last resort. (If the water is cloudy, double the dose.)

POWER TIP
Remember the ice cubes

Make sure your ice is made from water that has been boiled and cooled – freezing won't kill off bacteria and viruses.

Although pets are less susceptible than humans to infection from contaminated water, it is worth giving your pets treated water if possible, as they can be infected by waterborne parasitic protozoa like cryptosporidium.

Keeping water use low

In drought-ridden countries, one essential is to make sure you don't use up more water than you need to. Avoid being out in the hottest parts of the day – if you have to travel, do so in the early morning and evening onwards. Keep yourself covered up with light, breathing fabric, rather than exposing lots of skin to the sun, and stay in the shade whenever you can. This isn't just to avoid the risk of skin cancer, but is also an effective way to reduce the amount of water lost as evaporated sweat. Also:

❒ Keep your mouth closed – it's a good source of water vapour. This means limiting your chatting.

❒ Keep out of contact with anything hotter than the surrounding air. If the ground feels hot, don't lie on it.

❒ Avoid alcohol, coffee and tea. They are diuretics, removing water from the system.

❒ Move as little as you can. It's an excellent excuse to be lazy for a good reason.

❒ Don't smoke – again, this reduces water in the body.

❒ Minimize the amount you eat while it's very hot. This particularly applies to fatty foods. The digestive process uses up water.

Finding water

If your mains water fails, there are a number of alternatives.
Assume all sources (except bottled water) are contaminated
and treat them (see page 94). Bottled water is a quick and
easy (if expensive) short-term alternative to tap water.
Bottled water should be stored in the refrigerator once
opened and used within the period advised on the bottle.

Many houses have a cold-water tank. Even though this is
water from the mains it may have accumulated contamination in the tank, so still filter and boil it. You shouldn't
consider using the water in toilet cisterns and swimming
pools unless you can distil it, as both sources are likely to
have higher than usual chemical content.

A safer outdoor source is rain, which should be fine to
drink untreated, but it is usually worth filtering and boiling
collected rainwater in case your collection system has
attracted airborne pests. Use as wide a collector as possible –
plastic sheeting can be effective.

A spring or clear stream is the next best ground-based
source, and if all else fails, a garden pond – but don't use
water from a pond with dead animals in it, or with no green
plants growing around it, as the contamination may be
chemical and hard to remove. Assume any stream or pond is
infested with bacteria and parasites.

If you need to hunt for water outside, cast an eye over the topography. Water naturally runs to low points. Where would that be? Check out potential traps along the way that running water could be caught in.

In the winter, snow and ice form a natural water source, but don't try to melt snow or ice in your mouth – not only can it cause injuries, it tends to dehydrate. Ice has more water in it than snow, and takes less heat to melt, so is a natural first choice. Treat sea ice like seawater – it generally needs distilling to be made drinkable, although the older, blue sea ice has a lower salt content.

The living water pump

A large part of the working mechanism of a plant is engaged in searching out moisture in the ground and pumping it out – in effect, plants are living water pumps that can work for you: all you have to do is find a way to tap into that water.

If you place a clear plastic bag over a shrub, or the leafy branch of a tree, water that the plant has extracted from the ground and that is escaping as water vapour will be trapped in the bag. For a tree branch, use a large bag, tied around the branch, with one corner of the bag drooping to collect the condensation. For a shrub, use either a bag tied around the base of the shrub, or a plastic sheet, lifted above the top of the shrub with a stick. In either case, condensation will run down the underside of the 'roof' – make sure the plastic sheet or bag runs into a trench to collect the water before it can run out around the base of the plant.

Water from the Sun

The heat of the Sun can be used as a mechanism to extract water vapour from the ground which is then condensed for use. This is particularly effective when it is hot during the day but quite cold at night.

Dig a pit in the ground in a location that gets full sunlight most of the day, and place a large clear plastic sheet over it. The sheet should be weighted down around the edge to stop it from dropping right into the pit, but the centre of the sheet should also be weighted down with a stone so it droops towards the centre of the pit. Put a collecting bowl under the centre of the sheet at the bottom of the pit.

The plastic sheet acts like a greenhouse, trapping the heat and evaporating water, which then condenses when it comes into contact with the sheet, particularly as the air cools, running down to drip from the centre into your collecting vessel.

You can also use this mechanism to purify water by distillation. Place a bowl of contaminated water to one side in the base of the pit where it will catch the sunlight. Liquid will evaporate from the bowl, condense and drop back into the collecting vessel. As mentioned above, this won't remove volatile contaminants that boil and condense alongside the water, but will remove pretty well everything else.

Without a water closet

When water runs short, the toilet becomes an enemy, consuming large quantities of water in every flush. You can reduce the water consumption by putting a half-brick or a stone in the toilet's cistern, and only flushing when strictly necessary, but it may come to it that you have to abandon the flush toilet. There are a number of options, provided you have a garden and can construct your toilet well away from the house.

A simple urinal can be made by binding together a big bundle of rushes or straw (or just use a straw bale on end, partly dug into the ground). Put the bundle on the compost heap after a season. Alternatively, dig a hole, mostly fill it with medium to large stones, then put a cone of flexible material in the top, held in place with earth.

The simplest toilet for solid waste is an earth hole with a wooden seat above it (or used in a French-style standing position). Make sure the latrine is sited well away from the house, and that the prevailing run-away of water does not head from its position towards the house. There should be a shovel nearby with a pile of earth or sawdust – add a layer each time the toilet is used for solids. Have a separate urinal – this way, the toilet will smell less. Always cover the toilet when it is not in use to prevent flies getting access. Abandon it when there is still a good gap above to fill in with fresh soil and mark its position (perhaps with a non-edible plant) to avoid accidentally digging into it at a later date.

If you do stop using your normal toilets, the water will gradually evaporate from the bowl. When the level sinks low enough it will stop providing a seal – the S-bend stops foul air wafting up from the sewer. Don't flush the toilet, but do top it up with unwanted water (perhaps after you have washed in it). If water is terribly short, seal the top of the toilet bowl with plastic sheeting to reduce evaporation. (But remember it's there if you start to use it again!)

Shortage of water means you will probably be washing less, but don't skimp on hand washing as poor sanitation makes it easy for diseases that have been largely controlled in the western world, like cholera and typhoid, to make a reappearance. See if you can get hold of the waterless cleaning gels used in hospitals, to use between water washes.

⚡ POWER TIP
Garden cleansers

If soap runs short, it's quite difficult to make a true soap from animal fat and caustic soda (sodium hydroxide – beware, this can cause serious burns) or other alkalis like wood ash, but it is significantly easier to make substitutes from wild plants.

Probably the best of the plants for this is the aptly named soapwort (check a plant guide to see if this grows locally). In North America it is rather less helpfully called bouncing bet. This plant can be crushed between the hands in warm water, or boiled up in water to produce a soaping effect. A similar, if not quite so good, result can be had from the leaves of the horse chestnut (conker) tree, or from the leaves and flowers of the popular garden climber clematis. In North and Central America, yucca (often grown as a houseplant elsewhere) is a reasonable source: if the plant is plentiful, use the roots, which contain the most of the soap-like saponins; but if you have only a few plants, crush the leaves.

The store cupboard

Should global warming threaten stability, it will be essential to have a reasonable store of long-lasting food. Those whose parents or grandparents were alive during the Second World War, or similar conflicts, may have heard of government warnings against hoarding food, but this reflected both a requirement to avoid panic buying and rather less sophisticated food-preservation techniques in the last century.

Gradually build up a stock of key supplies: food, water and other household basics. If you have appropriate storage, you can build up supplies to last several months. Don't do this all at once, but take advantage of supermarket special offers to stock up on non-perishable and long-lasting goods.

POWER TIP

Special offers are a great investment

It's not unusual for supermarkets to offer 'buy one get one free' or a product that's briefly sold at half price. If this really is a short-term offer, it's not just a good way to stock up for emergencies, it's a great investment. We are encouraged to move our savings around to get the best rate of interest – but savings accounts rarely rival supermarket reductions. Imagine you could get a 10 per cent return on a savings account (significantly more than is available at the time of writing). If you invested £50, you would get £5 interest at the end of the year. But imagine you found £100 worth of goods at 50 per cent off that you would use anyway during that year. You have just invested £50 and got the equivalent of £50 interest – the money you didn't spend on the free goods.

Check the 'best before' dates – many canned and dried products have a significant lifetime. As you build up your stock, always use the oldest as you go along, so your store keeps up to date. Ideally don't buy long-life products you never normally use – that way you will continue to cycle through them, rather than find you've something five years past its sell-by date.

Store your long-term goods carefully. In the corner of a damp garage is not a good location, especially for packets. Make sure the food is stored in a cool, dry place, off the floor and ideally in closed cupboards or containers.

Apart from food and water, consider adding to your stock list basic hygiene essentials, disinfectant, medical supplies and batteries.

Beyond the can

Tinned food often keeps for several years (check the use-by date on the can), but there is a wide range of dried and other foods, some less common these days, that were used in the days before refrigerators and can be stored for a good length of time. Consider including the following in your stock:

❑ Rice – you may prefer the taste of brown rice, but white rice keeps considerably longer: up to three years as against under a year for brown.

❑ Wheat-based products – have a good shelf life if kept dry and cool. Keep plenty of pasta and sproutable whole grains (mix these with the likes of mung beans and lentils for a great, nutritious sprout salad). Pasta will last two to three years, and whole wheat grains as long as five. Oats aren't so long lived at around one year.

❏ Peas and beans – although we're now more used to these being fresh, frozen or canned, most peas and beans keep well dry and you can build up a good stock. Soya beans are a particularly effective source of nutrients. Remember that dried peas and beans need to be soaked overnight before use (ideally with a touch of sodium bicarbonate). Can last between five and ten years.

❏ Dried milk and eggs – these wartime staples come back into favour when times are hard, and keep much longer than the fresh equivalents. Dried eggs can last between five and ten years, and dried skimmed milk from three to five years.

The poisoned garden

The image of nature is that of a source of good things to eat, but we generally rely on someone who knows better to protect us from the aspects of nature that are less pleasant. If you have to become more self-reliant in finding food, you need to take on responsibility for ensuring that the food you eat is safe.

⚡ POWER TIP
Natural is not the same as safe

In part thanks to propaganda from people selling 'green' or 'organic' products it's easy to fall into the trap of equating natural with safe. Just because something you might eat (or drink) is natural does not make it safe. Many naturally occurring materials are poisonous – in fact most of the deadliest poisons in existence are found in nature. Do not eat or drink anything on the assumption that it's OK because it is natural – check it out first.

A few general principles can help a lot in avoiding poisonous foods.

First, don't assume that because we eat part of it, all of a plant is edible. In quite a number of cases we eat the only edible part of an otherwise poisonous plant. Common examples are the potato, tomato and rhubarb, where any other part of the plant is likely to be dangerous. In fact in all these examples, even the part we normally eat can prove a problem. Potatoes and tomatoes both belong to the same family as deadly nightshade. Green potatoes and any shoots can kill. Green tomatoes are poisonous unless cooked – similarly, neither potatoes nor rhubarb should be eaten uncooked.

Second, never eat something from the wild unless you are absolutely sure of what it is. You should be safe with a crop you have planted, but it's all too easy to confuse some nasty wild plants for similar-looking edible varieties. The pea family, for instance, is mostly poisonous. Both the common decorative tree laburnum and the sweet pea are dangerous to eat (as are uncooked kidney beans). Even some wild grasses have grains that are poisonous. Note, by the way, that edible grains can be poisoned by the fungal infection ergot. Look out for black blobby growths like little beans on the grain heads.

Perhaps the most obvious example of natural food danger is fungi. In principle, wild mushrooms make great food. In practice, there are so many deadly fungi, many of which aren't hugely different in appearance from the edible varieties, that it simply isn't worth the risk unless you've an expert on hand.

 This really is worth emphasizing. Not only are some fungi deadly poisonous, there is no antidote. Only experts should give the OK to eat a wild fungus. Don't rely on identification guides – the real thing is always subtly different.

Is it edible?

In really dire straits, you can use a testing sequence to reduce the chances of eating a poisonous plant. With an unknown plant, take the following approach:

❐ Avoid all fungi (mushrooms) – the testing strategy below will not work with fungi. Many poisonous fungi don't give any indication until it's too late.

❐ Avoid danger markers – don't eat plants with milky sap (except dandelions), mature bracken (the young, tight-coiled 'fiddle heads' are fine, though), plants with bell-shaped flowers, fruits that divide naturally into five segments, red plants and those with red seeds, and plants with tiny barbs that mean the plant or seed tends to catch on material. Unless you planted it and know it's an onion or garlic, avoid bulbs, most of which are poisonous.

❐ Take a good look – does the plant look old, or decayed, or partly eaten? If so, discard it. If you know anything about plants, try to identify it.

❐ Crush a small part of it and sniff it – does it smell of bitter almonds or peaches? If so, discard it, it may contain a cyanide-related poison.

❐ Squeeze some juice from it and rub on a delicate part of your skin – if it causes irritation, discard it. Allow a few minutes before moving on.

❐ Put a small piece on your lips, without taking it into your mouth – if there is any feeling of irritation or discomfort, discard it. Allow a few seconds before removing and moving on.

❐ Put a small piece in the corner of your mouth without swallowing – if there is any feeling of irritation or discomfort, discard it. Allow a few seconds before removing and moving on.

❏ Put a small piece on the tip of your tongue, without swallowing – if there is any feeling of irritation or discomfort, discard it. Allow a few seconds before spitting out and moving on.

❏ Put a small piece under your tongue – if there is any feeling of irritation or discomfort, discard it. Allow a few seconds before spitting out and moving on.

❏ Swallow a small amount – watch out for reactions. Wait for six hours before moving on. To be sure of the cause of any problem don't eat or drink anything else in this time.

❏ First proper eating – again, allow a six-hour period before assuming that the plant is edible.

It's a slow, painstaking process, but your survival is something that you want to take pains with.

Outdoor eating

Of course, not everything out there is poisonous. Even those with minimal country lore can recognize wild blackberries (brambles) in the hedgerow, and there are many more edible delights that can be collected, particularly during the summer months. A simple rule of thumb (which rules out all fungi) is to collect only food plants that you can positively identify as safe to eat. Nettles and young beech leaves, for instance, make good cooked greens, while young hawthorn leaves and dandelions are good contributors to salads. Also don't forget nuts in the autumn too (but don't confuse the poisonous horse chestnut with the edible sweet chestnut).

This isn't the place for a detailed guide to your local edible plants – get a good illustrated guide.

Worm sandwiches

The living, breathing nutrition in the garden that's easiest to catch comes in forms we are least likely to eat – but when things are tough, it's worth overcoming that natural reaction and facing up to the wriggly delights that are an awful lot easier to lay your hands on (and, to be honest, to kill) than a fish or a mammal.

Perhaps best of the lot is the common earthworm, which is easy to find with a garden fork in practically any soil. The only problem with worms is their eating habits – they operate by passing soil through their body and extracting nutrients, so when first dug up, a worm has an unpleasant core of earth. The simplest way to clean them is to rinse them well in clean water, put them in a box with a fine grating on the bottom, away from the soil, for twenty-four hours, then give them another good wash. Alternatively, the less squeamish can try squeezing the worms gently but firmly along towards the tail to expel the dirt, but this may waste a few worms until you get the hang of it. Worms can be eaten raw (though they should be sterilized because of bacteria from the soil), but it's best either to dry the worms first or fry them up to make them more palatable.

As the French have told us for years, snails are edible (and really not unpleasant if cooked correctly). Again, put them in a box with a grating on the bottom to cleanse themselves. Give them a good rinse of water and leave for twenty-four hours – repeat this, so they have spent three days in the box. Don't put them in salt as the French used to – this has no advantage and is unnecessarily cruel. Make sure your snails are still alive – discard dead ones. Then drop them into boiling water and keep them at the boil for three minutes. Drain the snails and rinse in cold water for several minutes, then remove them from their shells. (A thin-bladed knife or knitting needle should do the job – use a twisting action.) Wash the snails three times in a vinegar/water mix, then drain well and braise for around thirty minutes. Although slugs are traditionally not eaten, from the point of view of edibility they are no different from snails. It's best to avoid sea snails and tropical snails with brightly coloured shells which may be poisonous.

As viewers of various reality TV shows are aware, there is a wide range of creepy-crawlies that are also edible (if often harder to catch and having less nutritional value than worms and snails). Grasshoppers, locusts, crickets and the like normally have more body mass than most insects. Remove the heads, wings and legs before cooking – they can be boiled or roasted. Ants are surprisingly high in food value if you get enough, though they do use the unpleasant formic acid (used to descale kettles) as a sting, so need to be boiled for around ten minutes to make sure this is broken down. Grubs are often great in food value too, though it's best to have some form of 'bush tucker' guide to make sure you are getting edible ones, and avoid anything highly coloured, whether a fully grown insect or a grub. As with worms, starve grubs and wash them well before cooking to clear out any dirt.

Leather on the menu

Ever since Charlie Chaplin ate his belt in the movies there has been a myth that eating leather is not much different from eating meat. It is different. Leather is chemically treated animal skin, so it didn't have huge food value to begin with and has considerably less after it has been soaked in some pretty foul substances (a traditional ingredient in tanning was dog excrement). If you really do want to eat leather – and it's essential to make sure the item *is* leather, not imitation – it will need to be boiled for a considerable time with a couple of changes of water to make it palatable. Even then it's not a great food source and should very much be regarded as the last resort.

Fish without a packet

If you are relying on what you can lay your hands on, fish can be a great source of protein, are easier to catch than most animals, and are easier to kill than something cute and furry. But you may not be familiar with the simple steps of preparing a fish for eating. Make sure you've got an edible variety (get yourself an appropriate illustrated guide).

❐ Cut a slit along the base of the fish from behind the gills with a sharp knife to just before the anal opening. Pull out the innards carefully.

❐ Wash both inside and out.

❐ Ideally leave the fish for a period of five to ten hours, which will make it easier to finish the preparation process.

❐ Take off the tail and fins with a sharp knife.

❐ Cut around the fish behind the gills, cutting down to the bone but not through.

❏ Take the head in your hand with your thumb resting on the spine and peel the spine away from the flesh. You may find it helpful after starting the fish to turn it over. If done smoothly, most of the ribs should come out with the spine, but as always with fish, beware of bones that are left behind when eating it.

Bunny for brunch

Definitely not for vegetarians. Most mammals are good sources of meat. It's not possible to cover them all here – get an outdoor survival guide – but after fish, rabbits (or similar animals like guinea pigs) are often the next most readily available source of meat. Do beware of a diet that is too dependent on rabbits, though. Although they make excellent eating (in fact rabbits were introduced as food animals to many parts of the world where they are now considered native), rabbit alone is not a sufficient diet. It lacks fat (making it great for weight loss) and minerals, both of which are needed in small amounts. Canadian trappers have starved to death on a plenteous diet of rabbits.

Rabbit, like game, is often hung for a few days in a cool, airy location to tenderize the meat and bring out the flavour. This is not essential. Many wild rabbits suffer from the disease myxomatosis, which results in inflamed mucous membranes. This doesn't make the animal inedible, though as a general rule healthy animals should be selected in preference to those with any obvious disease.

Again, if you regularly buy rabbit from the supermarket you will be used to getting an animal already skinned and gutted, but the process isn't too painful.

❏ Hold the rabbit by the ears with one hand and with the other take hold of the stomach: squeeze and press down firmly to expel any urine.

❏ Pinch and lift the fur at the stomach and cut through the skin with a sharp knife, slicing from the middle of the stomach up to the sternum and back towards the anus. When you have cut a line, pull the skin away from the stomach wall.

❏ Make a similar cut through the stomach wall, but being careful not to disturb the internal organs.

❏ Hold the carcass belly-down and shake it to expel the internal organs – put your hand in and clean out the cavity.

❏ Cut away the rest of the skin, cutting between skin and flesh so it comes away whole. When cut around, pull the skin away from the flesh.

❏ Remove head and feet. The rest of the carcass is now ready for cutting up and cooking. Rabbit makes a great stew.

Vitamins and minerals

When food supplies become limited you may find yourself in danger of running low on vitamins and minerals. As part of your stocking up of essentials, get in a supply of multi-vitamins – this is one of the rare occasions supplements will definitely be of value. Unfortunately, however, these won't keep for ever, so you will either have to throw them away occasionally or use them, then replace them with new stock.

> ⚠ *Some vitamins are taken in very small doses. Note the difference between a milligram (mg) – 1,000th of a gram – and a microgram (µg) – 1,000,000th of a gram.*

Some key deficiencies to watch out for are:

❏ Salt – it might seem strange that it's dangerous to drink seawater, when we all need a small amount of salt on a regular basis. The important words here are 'small amount'. An adult needs around 1 gram of salt a day (about half a level teaspoon), and shouldn't exceed 3 grams. A glass of seawater contains around 8 grams of salt – and anyway, you shouldn't take all your salt in one go. Make sure you are getting some salt (if necessary, boil off some sea water to produce it), but keep the amounts small. The blood of animals is a good source of minerals, including salt.

❐ Vitamin A – helps with eyesight and growth, while a deficiency can produce night blindness. Vitamin A is present in liver, cod-liver oil, carrots, green leafy vegetables and egg yolks. Some margarines are enriched with vitamin A. The recommended daily amount is 750 to 900 micrograms (technically, retinol equivalent micrograms). Excess vitamin A is harmful over a sustained period, which is why liver should not be eaten more than once a week, as it contains a massive dose.

❐ Vitamin B_1 (thiamine) – essential for your nervous system and heart; too little vitamin B_1 and you can lose concentration, become confused and exhausted, lose your balance easily and get a tingling sensation in the toes. The extreme form of the deficiency is beri-beri. Unlike for vitamin A, there is no real maximum safe intake, as excess B_1 just passes through the body. You are aiming for 0.8 to 1.2 milligrams a day. Apart from supplements and vitamins added to breakfast cereals, you will find vitamin B_1 in rice, milk, liver, pork, peanuts, yeast and yeast products (particularly good), and wholemeal bread and cereals. High temperatures, alcohol and coffee all destroy B_1.

❐ Vitamin B_2 (riboflavin) – helps your skin, nails, hair and eyesight; a lack of B_2 results in itchy skin and eyes. Like B_1 there is no real maximum dosage, as excess just passes through the body. You are aiming for 1 to 1.7 milligrams a day. Apart from supplements and vitamins added to breakfast cereals, you will find B_2 in cheese and milk, liver, green leafy vegetables and fish. Too much light makes B$_2$ break down (as does alcohol), so it's best to keep foods containing B$_2$ in the dark.

❏ Vitamin B$_6$ (pyridoxine) – useful for the skin and the nervous system; a shortage of B$_6$ leads to skin problems, sleeplessness, irritability, and at its extreme to convulsions and anaemia. A recommended daily dose is around 1 to 2 milligrams. Apart from supplements, B$_6$ is found in bananas, beans, chicken, fish and pork. Heavy cooking and alcohol tend to destroy B$_6$, as does the hormone oestrogen present in the contraceptive pill.

❏ Vitamin B$_{12}$ (cobalamin) – essential for supporting the formation of red blood cells and nerves; a lack of vitamin B$_{12}$ results in anaemia and nerve damage. A recommended daily dose is around 2 milligrams. Apart from supplements, most of the sources of B$_{12}$ are animal-based – fish, meat and dairy. As these may be in short supply when fending for yourself (or if you are vegan), you may need to look for supplements. However, this is only a long-term issue, as the liver can store B$_{12}$ for up to five years. It is destroyed by alkalis (such as baking powder).

❏ Vitamin C – there is a lot of confusion over the ideal vitamin C intake. There's no doubt we need some – without it, the human body is susceptible to the horrible disorder scurvy, once common on ocean-going ships – but exactly how much we need is widely disputed. The basic recom-mended dose for adults advised in many countries is around 60 to 95 milligrams per day. Others recommend 200 to 400 milligrams, while those who believe in the thera-peutic benefits of vitamin C argue for as much as 3 to 5 grams a day, though there is a concern that doses of over 2 grams may cause indigestion or diarrhoea. If you are short of vitamin C, before the full-scale outbreak of scurvy you are likely to suffer from bleeding gums, wounds that heal more slowly than usual and tiredness. There are plenty of supplements containing vitamin C, though these should only be necessary if you can't get to the rich sources: citrus fruits, kiwi fruit (especially good), berries, tomatoes, potatoes, peppers and green leafy vegetables (including cauliflower, which is a mutant cabbage). Raw is best for vitamin C – cooking removes around 25 per cent.

❏ Vitamin D – essential for strong bones and teeth; a shortage can lead to osteoporosis, rickets, and increased chance of a number of cancers. We need around 5 to 10 micrograms a day. The best source of vitamin D is sunlight on the skin (see page 142), but it is also found in cod-liver oil and oily fish, and is often added as a supplement to milk and margarine.

❏ Vitamin E – an important antioxidant that removes toxins from the body; if vitamin E is in short supply there is a tendency towards physical weakness and infertility. The recommended dose is around 7 to 10 milligrams a day. Vitamin E is found in broccoli, green leafy vegetables, soya, vegetable oils, nuts and eggs. It can be destroyed by heat, cold, chlorine and oxygen.

❏ Vitamin K – used in the formation of special blood factors which help with clotting and reduce the risk of haemorrhage or bleeding, as well as contributing to bone and kidney functions. Around half our vitamin K requirement is produced by bacteria in the gut. The remaining requirement is around 70 to 140 micrograms a day (roughly 2 micrograms per kilogram of body weight). Vitamin K is found in green leafy vegetables, wheat and some meats.

❏ Folic acid (folacin) – important for production of red blood cells and essential during the first three months of pregnancy; a shortage of folic acid can cause birth defects in babies and anaemia in adults. The recommended daily amount is 200 micrograms, but pregnant women are recommended to double this. Found in carrots, liver, beans, green leafy vegetables, whole wheat and yeast products, and often added as a supplement to breakfast cereals, folic acid is destroyed by excess heat, oxygen, sunlight and oestrogen.

❑ Niacin – a significant requirement for good skin and the nervous system; shortages of niacin can lead to weakness, loss of appetite, pellagra displaying as dermatitis in skin exposed to the sun, diarrhoea and in extreme cases dementia. A daily intake of 10 to 20 milligrams is recommended. Niacin is found in wholegrain products, fish, most meat, peanuts and sesame seeds, and is often added as a supplement to breakfast cereals.

❑ Minerals – as well as salt, there are a number of minerals essential to the diet, including calcium, magnesium, iron and zinc. Our main source of calcium is dairy products, so vegans or those unable to obtain dairy should be careful that suitable supplements are taken. Soya does contain some calcium, but not as much as dairy, though commercial soya products often have extra calcium added. Magnesium deficiency is rare as most foods are a reasonable source: it often results from prolonged vomiting/diarrhoea, or excess alcohol consumption. Although iron is present in a good number of fruits and vegetables it is more readily accepted from meats (so Popeye's spinach wasn't as good as he thought). Zinc, similarly, is best absorbed from meat and eggs. Vegetables, and whole grains in particular, don't provide zinc very well. Zinc deficiency can lead to delayed healing, skin irritation and loss of the sense of taste.

See page 128 for examples of vitamin content in typical foods.

Living off the land

It's easy to have a rose-tinted, 'good life' view of living off the land. In reality, subsistence farming is back-breaking, low-return work. It will suit some, but isn't for everyone.

⚡ POWER TIP
Sprout your own

Even if your only garden is a dead pot-plant on the windowsill, you can do one bit of agriculture in your home. Sprouting seeds and pulses will give you fresh, healthy food with little effort and without the need to touch compost or fertilizer. You can buy a purpose-built sprouter, or simply use a large glass jar (1 litre or larger) and a piece of fine-mesh cloth such as a (clean) pair of tights or muslin.

Use specifically designated sprouting seeds (some seeds are poisonous). Rinse them off and place them in your jar that you have half filled with warm water (best to boil the water and let it cool to remove any chlorine). Cover the jar with your cloth and use an elastic band to fix it in place. Soak the seeds overnight (or as directed on the pack), then drain them and rinse with cool water that has been dechlorinated by boiling first, or by leaving it to stand for several hours.

Keep the seeds in the jar, warm and dark (an airing cupboard is good), while they germinate. After two to three days of rinsing and draining two to three times a day you will have a harvest. If you prefer your sprouts to be on the green side, give them a little light once they have germinated. Check the pack for advice on eating: some suggest that certain bean sprouts should not be eaten raw in case there are poisons left from the beans.

To be totally self-sufficient, you need a minimum of around 5 acres (2 hectares) of productive land. But this doesn't mean that you can't do anything with less than that. Even the smallest city back yard can help bolster your food supplies and make it more practical to cope with shortages. Possible ways to use the area you have available are:

❐ The city yard – in a small area you can produce some basics. If you have a south-facing wall, this is ideal to grow delicate crops against. Preparing the depth of soil needed for effective growing (an absolute minimum of 0.5 metres) can be difficult in a city environment. If necessary, you can build raised beds with brick or wooden railway sleeper surrounds, reducing the amount of depth you need to get below existing ground level. Try growing root crops that don't take up too much room, like carrots, or crops that do most of their work above ground like many of the pulses – beans and peas. Tomatoes love the shelter of a town garden, require little depth of soil (growbags are fine), and have health benefits too. To reduce your dependence on buying in fertilizer, a composting bin (both for garden waste and kitchen vegetable waste) is a good investment.

❏ The suburban garden – won't provide enough land for self-sufficiency, but is one step on towards surviving on your land. Although the approach is broadly the same as for the city yard, you will have the advantage of much more workable soil and less need to consider raised beds and other work-arounds. You can also extend your crops to more space-eating possibilities like potatoes. Cabbage, cauliflower, Brussels sprouts and related crops are also more practical here and give good food value. If you don't want to turn your entire garden into one enormous vegetable patch, you can, of course, split off a part of it, perhaps divided by hedging or even fruit trees grown in the flat 'espalier' style against wires. Alternatively, in some countries you can rent plots of land ('allotments' in the UK) purely for growing vegetables and fruit.

❏ A 1-acre plot (0.2 hectare) – with this much land you have reached a farm in miniature. You may be able to grow enough to support a modest number of people, but if you decide to have animals, you won't be able to grow enough to feed them, so will have to buy food in. Realistically you are already at the stage that at least one member of the family has a full-time job of looking after your small-holding; this is no longer weekend and evening work.

Another major consideration with livestock (particularly those requiring milking) is that you won't be able to get away from the place even for a couple of days unless you can get in someone to look after the animals for you. Perhaps the first livestock to consider are egg-laying hens, requiring very limited maintenance (as long as you keep predators like foxes out), and producing food that even the squeamish can cope with. You could also consider a cow or goats or perhaps pigs (though look at pigs only if you are prepared to have them butchered and to eat them). Having animals (other than chickens) does mean giving over a fair amount of land to grazing – a cow will need half an acre in all, occupying a quarter of it at any one time in rotation. If you have a cow it will need winter accommodation (this in particular is where the requirement to buy in feed comes in).

❒ A 5-acre plot (2 hectares) – here at last is true self-sufficiency. Of course there will be things you need to buy in, whether equipment or luxuries like foodstuffs that can't be grown in your area, but you should be able to feed a good-sized family with more left over. By this stage, like it or not, you have become farmers. Keeping this set-up going is a job for two people. (If you think that modern farms are much more efficient than this, bear in mind you may have to manage without electricity or fuel.) Your animal holdings could increase proportionately. If there is doubt about the availability of diesel (or your ability to buy it), you may want to include a horse to provide power for the essential groundwork.

Befriending a farmer

Farmers have a yo-yo relationship with the press. At one time they were rosy-cheeked sons of the soil who could do no wrong. This attitude still prevails in some countries where there remains a strong tradition of peasant farming, but in many western countries, as subsidies were introduced to ensure that local growing capability was maintained, farmers began to be regarded as greedy parasites who took all they could from the authorities while giving little concern to the consumer, an image reinforced by apparent attempts to cover up BSE and other health scares. Later still, as barriers to trade were relaxed and as supermarkets became more powerful, farmers became the underdogs, undertaking back-breaking work to earn a minimal return.

However farmers are portrayed, they are a relatively rare breed these days, which is a pity as they have a real advantage in times of food and water shortages. Most farms will have more stored water than an ordinary home, and at the very least are likely to have one food crop. Many farmers still like to dabble with other farmyard activities, so despite running, for example, a dairy farm, may also keep chickens.

When things are short it's too late, so if you haven't already got a farmer friend, now's the time to cultivate one. If you live in the country, this is relatively easy. Ask around the school playground or at the village shop. Get to know at least one of your local farmers. If you can, do them a few favours. Make friends. When things get tough, you'll consider every minute of getting to know the farmer worthwhile. What's more, you'll feel more part of the countryside, rather than being a transplanted townie.

If you live in a city, this is harder, but not impossible. Farmers' markets are spreading into towns and cities, so you can start by buying their produce. Many direct-sell farms also give you the chance to come and see the farm in action – take this up. You could try holidaying on farms and seeing if you can strike up a friendship, or rent a cottage in the countryside and regularly visit. (It's best not to buy a second home, even if you can afford it, as this can build up resentment, rather than have the desired effect of getting closer to the local farming community. Second homes mean fewer places to live for the locals, and specifically the farmers' children.)

It may seem extreme, but it is worth city dwellers considering moving to the country. After all, many people are prepared to move to get their children into a good school – how about moving to get them into a future? Not only will you increase your chances of getting close to the land, you will have a better quality of life – and with modern technology it's perfectly feasible to live in the country but keep connected to a city business. Of course it's not for everyone, and it does make it harder to get to the entertainments of civilization, but many would argue it's not a bad price to pay.

The whole idea of befriending a farmer may seem horribly contrived and forced. But in any situation where we first meet people things are artificial – it's only as we get to know people that we develop a real relationship. This is just providing a kickstart in the way you might by joining a local club. In essence, what's recommended here is building up social credit. Establishing a community of people who help each other out. And that can't be a bad thing. Of course, you may find you just don't get on with your potential farmer friends. Fine – there's no suggestion you should pretend to get on when you don't. But at least consider the possibility.

Town scavenging

When things become difficult, towns and cities are less easy places to live than the countryside. Although scavengers like foxes have become adept at town life, they do so by making use of resources few humans would consider.

POWER TIP
Windowsills and window boxes

Even if you have no outside space whatsoever, you can grow some edible plants in the home. Tomatoes flourish, either in pots or better (if messier) in a growbag on a sunny windowsill. Most of the smaller vegetables and pretty well any herb can be cultivated indoors. If you have a balcony or access to a roof space, then you can do even more. It's not going to feed your family, but every little helps.

If you need to stay in the town when food or water is scarce, there are limited options available. For water, look to all available natural sources. Is there a river or stream in the town? Make use of rainwater, collecting water from downspouts and using collecting containers on balconies or in yards. Solar stills (see page 99) can be as effective in the city as in the countryside. Look for artificial sources too. Air-conditioning can produce a considerable amount of water from the air – if your building has air-con, where does the water go? Are there water tanks in the building or the area?

> ⚠️ *Bear in mind water tanks linked to air-conditioning systems and the like can be breeding grounds for bacteria – treat them as the urban equivalent of pond water (see page 98). Because of the risk from legionella (Legionnaire's Disease), spread by water spray, only approach such water sources masked.*

You are unlikely to find much food growing in the city, though most parks will have some edible plants. If things are truly desperate, then like the fox you will have to be prepared to hunt around the bins and dumpsters. Obviously any food that has been discarded must be assumed to be tainted in some way and at the very least will need serious cooking to ensure any bacteria are killed. You will also want to cut away any obviously diseased, part-eaten or decayed material. A particularly good source may be round the back of supermarkets, looking for food that is recently past sell-by dates, and behind restaurants, bakeries and caterers, where food is often thrown out because it is not consumed within a short period of being cooked. You will probably also find part-drunk bottles and fast-food cups of fluids. Be wary – what it says on the bottle isn't necessarily what's in it – but if all else fails, with suitable treatment this can be a good source of water in a troubled city.

If you do resort to fishing in the rubbish, take some simple precautions. First you'll need gloves. Wear a pair of gardening gloves or other thick gloves to avoid getting cut by open cans and other sharp objects. The best sort are the industrial gloves with a rubber coating on the outside – otherwise you may wish to add an outer layer of kitchen gloves to stop the glop and slime soaking into the fabric. These gloves will make handling things harder, but are essential to reduce the risk. Take a pair of plastic sacks, one inside the other to add strength, to carry anything you collect. Wear an old coat, or some other outer wear you don't mind getting splashed. And whatever you do, when you've been handling the rubbish, resist any urge to touch your eyes, nose or mouth until you've put the gloves aside and had a good wash of your hands with disinfectant cleanser.

Some individuals may resort to looting if there is a major breakdown in law and order. Bear in mind that looting is treated very seriously, and looters can be (and often are) killed if martial law has been established. Of course if it's life or death and there is no alternative it may be necessary to break into a shop to access food or water, but bear in mind the risk. Try instead to negotiate with the owner, if present, and consider finding some way to repay what you have taken.

What you need to eat and drink per day

As we've already seen you will need between one litre (sitting in the shade, inactive) and five litres of water a day. Food requirements are more flexible, as most of us can live on reserves for several weeks if necessary. The guideline daily intake is around 1,800 calories for children aged between five and ten, 2,000 calories for women and 2,500 for men – this assumes average activity. If engaging in heavy manual work, you need more; if largely sedentary, less.

This isn't the place to list the calorie content of every food – there are plenty of calorie-counter books out there. One small point: the conventional distribution of food between our meals is more about social eating than about best use of the food (later meals tend to be more social events). In purely practical terms, it is better to have a large breakfast, then a light lunch, light tea and a snack mid-evening than it is to load up with food in the evening.

Resources

Containers

When it becomes necessary to store water, it's time to take stock of your appropriate containers. If there's any chance at all of water being cut off it doesn't do any harm to keep a few sealed containers of water stored away. If you don't have proper water carriers, large fizzy drinks bottles are good. Try to have at least 5 litres per household member stored away all the time (though the more the better).

When you receive a warning that you are going to lose your water supply, be more creative with your containers. A bath is great because it's so easy to fill (see page 92 about covering it). Pans are good, especially if they have lids. Look for any watertight containers, preferably with a lid or top that can keep out flies and pollution. Think before you throw away any containers that will hold more than a litre – stockpile them against water shortage.

Note that water in containers will lose its protection from chlorination over a few days. Even if it's drinkable tap water, boil it before use if it has been sitting in a container for more than twenty-four hours.

Plastic sheeting

 Keep plastic sheeting and plastic bags away from small children to avoid accidental suffocation.

Plastic sheeting may not be very eco-friendly, but it is certainly climate-change survival friendly. Opaque sheeting – black, for instance – is good for covering large water containers like a bath, or for lining a water-collection or storage pond. Clear plastic sheeting is essential for a solar still or to collect moisture from plants. If you come by any large plastic sheets, put them in storage, not in the rubbish. Folded they will take up very little room. If you don't have anything else, black binliners can be effective as opaque sheeting for covering a container, especially if they are the thick heavy-duty variety, though they won't be strong enough to line a storage pond.

Paddling pools and swimming pools

Any systems for storing a large amount of liquid can be hijacked in an emergency to hold water for household use. A typical domestic swimming pool may hold 40,000–50,000 litres, enough to keep a good number of people in water for a year or more. We may not all run to swimming pools, which have the advantage of usually being full of water already, but many of us have (or can cheaply get hold of) paddling pools. The capacity might not be so great, but they can still hold a good number of litres. Get one with a lid, or make a cover to keep the water as clean as possible. Covered plastic sandpits make an excellent alternative.

Reference books

A few extra books can make a very valuable resource to help cope with finding food and drink, particularly a manual of the edible plants of your part of the world. When food is short it's also useful to have a nutritional guide so that you can deal efficiently with rations.

Camping stove

Strangely, this is an essential resource for water, rather than food. It's relatively easy to improvise a way of filtering water – through fabric, for instance – but it's hard to improvise a way of boiling it, if you haven't got access to power. If all else fails you can use a fire, provided you can get one alight, but a camping stove is a cheap, quick way of boiling water so you've something safe to drink.

 Don't use a camping stove indoors – they aren't designed to be used in confined spaces and may give off a significant amount of toxic fumes. Make sure that wherever it is used is well ventilated.

CHECKLISTS

Water purification kit

If you can, buy a professionally made water purification kit (filter-based, not chemical), but in the absence of this, here's the basics.

Filter .... ❏
Essential to get debris out of polluted water. Ideally use a coarse-grain filter first – fine-weave cloth (like muslin) or a coffee filter is good. Then, if possible use a charcoal filter like those in a household water filter. This will remove some of the smaller particles that would get through the coarse filter.

Pan (not aluminium) or kettle .... ❏
Filtering isn't enough – you need to boil the water for three minutes too.

Camping stove .... ❏
It's fine to use your usual cooker to heat the water, but having a camping stove in case the power or gas is off makes it more likely that you will be able to get clean water.

Distillation kit .... ❏
As an alternative to the pan and stove, a still will produce pure water, even removing dissolved chemicals (unless they have a similar boiling point to that of water).

Covered storage .... ❏
You need to leave the water to cool somewhere that is protected from flies and airborne matter.

Clear plastic sheet .... ❏
To build a solar still. This will purify better than filtration, so is ideal for treating particularly unpleasant water.

Store-cupboard basics

What you choose to store away is up to you, but this checklist gives an idea of the quantities required to keep one person going for a year. This assumes that you haven't got access to a freezer, but you can make the store more interesting if you can include frozen goods – check your freezer for details of how long you can keep different foods in it.

Rice, pasta and flour – 150 kg .... ❏
Stick to white rice, as brown doesn't last anywhere near as long. Don't keep the flour in paper bags, use stronger sealed containers.

Dry legumes – 35 kg .... ❏
Peas, beans and lentils in dry form will last several years and provide a good protein input to your diet. You might want to have some canned legumes as well for instant use.

Dried dairy products – 25 kg .... ❏
Dried skimmed milk and eggs may not be as appetizing as the real thing, but they store well, and are better than going without.

Canned meat – 20 kg .... ❏
Canned meat will typically keep at least a year. If you are vegetarian there are dried meat-substitute products you can use. You will need significantly less weight.

Sprouting seeds – 20 kg .... ❏
Sprouting seeds are doubly valuable – you get fresh food and they're nourishing and tasty too. Expect these to be a regular part of your diet. See page 117.

Fats and oils – 20 kg .... ❏
As well as cooking oils, this includes spreads for bread (butter or margarine) etc, though these will require refrigeration.

Sweeteners – 15 kg .... ❏
Sugar and honey to taste.

Drinking materials – 15 kg ....❏
Depending on your preference, tea, coffee, drinking chocolate etc.

Dried fruit and vegetables – 10 kg ....❏
Some produce – potatoes and apples for instance – can be kept several months if stored from freshly grown in a dark, cool, dry environment, but to last longer you will need dried.

Assorted dry goods ....❏
You will need salt, yeast to make bread, baking powder and any vitamins and mineral substitutes you might need (see page 112)

Assorted other canned/bottled foods ....❏
Some preserves will last a surprisingly long time and help to make your store more interesting.

Vitamin and mineral content

Pages 112 detail the key deficiencies to avoid. Here are figures for typical content of important vitamins and minerals in some sample foods. 'Percentage of RDA' is the percentage of recommended daily amount – i.e. you are recommended to have a total intake of 100 per cent per day. In most cases, exceeding the recommended daily amount won't do any harm; the vitamin that's most risky in large doses is vitamin A, which is why it's recommended that you don't eat liver more than once a week, and don't eat the livers of some animals (dogs, for instance) at all, as they can accumulate a deadly quantity of vitamin A.

Vitamin A	Content (µg)	% of adult RDA
Cheese (Cheddar): 25 g portion	78	10
Liver (fried): 100 g portion	26,780	3,570
Margarine: 10 g	58	8

Vitamin B_1	Content (mg)	% of adult RDA
Bacon: 2 rashers	0.1	10
Cornflakes: 30 g portion	0.27	27
Marmite/Vegemite: 2 g	0.08	8
Peanuts: 25 g	0.15	15

Vitamin B_2	Content (mg)	% of adult RDA
Cheese (Cheddar): 25 g portion	0.125	8
Cornflakes: 30 g portion	0.4	25
Liver (fried): 100 g portion	4.4	275
Marmite/Vegemite: 2 g	0.32	20

Vitamin B_6	Content (mg)	% of adult RDA
Banana: 100 g	0.5	50
Bread (wholemeal): 1 slice	0.025	2.5
Liver (fried): 100 g portion	0.5	50
Marmite/Vegemite: 2 g	0.06	6
Potatoes (boiled): 100 g	0.2	20
Salmon (steamed): 100 g portion	0.5	50
Walnuts: 20 g	0.14	14

Vitamin B$_{12}$	Content (mg)	% of adult RDA
Cheese (Cheddar): 25 g	0.4	20
Egg (boiled): 1	0.8	40
Liver (fried): 100 g portion	81	4,050
Oysters (raw): 100 g	15	750
Pork chop (grilled): 100 g	1	50
Stewing steak (stewed): 100 g	2	100

Vitamin C	Content (mg)	% of adult RDA
Broccoli (boiled): 100 g portion	34	57
Cabbage (raw): 50 g portion	30	50
Kiwi fruit: 1	35	58
Orange juice: 200 ml glass	78	130
Pepper (red): 1/4	56	93
Potatoes (boiled): 100 g portion	9	15
Strawberries (fresh): 100 g	60	100
Tomato: 1	14	23

Vitamin D	Content (µg)	% of adult RDA
Cod-liver oil: 2 teaspoons (10 ml)	5	50

Oily fish: 100 g	5	50

Sunlight on skin provides more than enough in appropriate sun: see page 142.

Vitamin E	Content (mg)	% of adult RDA
Blackberries: 100 g	3.5	39
Egg (boiled): 1	0.9	10
Hazelnuts: 15 g	3.15	35

Vegetable oils high in polyunsaturates tend to be rich in vitamin E.

Vitamin K	Content (µg)	% of adult RDA
Broccoli: 100 g	100	100
Cabbage (raw): 50 g	50	50
Spinach: 60 g	144	144

Folic acid	Content (µg)	% of adult RDA
Almonds: 15 g	14.4	7
Bread (wholemeal): 1 slice	10	5
Broccoli: 100 g	110	55
Liver: 100 g portion	240	120
Orange: 1	36	18
Spinach: 60 g	84	42

Niacin	Content (mg)	% of adult RDA
Bread (wholemeal): 1 slice	0.6	4
Cornflakes: 30 g	2.7	19
Beef mince (stewed): 100 g	4.5	32
Peanuts (salted): 25 g	4	29
Salmon (steamed): 100 g portion	7	50

FOUR

WILD WEATHER

It should come as no surprise that climate change is bringing greater extremes in the weather. Coping with hotter summers, droughts, dramatic storms and, perversely, colder winters is likely to be a reality for all of us.

What to do ...

Drought, flood, freeze and storm

Humans are surprisingly poor at adapting to variations in temperature. Unfortunately, climate change means that many of us will have to cope with more intense summer heat. The World Health Organization estimates that climate change has already claimed 150,000 lives a year over the last thirty years – and the death rate is growing rapidly. In one recent hot summer (2003), over 35,000 died in Europe alone from direct heat-related causes. Many human diseases are linked to climate fluctuations. Heatwaves induce heart attacks and dramatically boost respiratory illnesses. To make matters worse, sustained cold weather will also reap a grim crop of lives.

Putting unusually cold weather alongside heatwaves is not as bizarre as it seems. Climate change can also mean that winters get harsher, and should natural climate-control systems like the Gulf Stream fail, as some scientists have predicted, parts of the globe will suffer much lower temperatures. As we have seen, the Gulf Stream is part of an ocean conveyor system that brings warm water to areas like Northern Europe. The reality of ocean conveyor shutdown, if it happens, will not be as dramatic or as speedy as in *The Day After Tomorrow*. Even so, losing all or part of the conveyor would make cold winters even harsher for affected areas.

Just as climate change has the potential to bring unusual heat and unexpected cold, so it can be responsible for both drought and flood. Some countries, like the UK, are experiencing drier summers alongside more instances of heavy downpours of rain. As drought spreads, flooding incidents are getting worse. In 2002, central Europe was devastated by dramatic floods that left a clean-up bill of £1 billion in Prague alone. Floods that might have been expected once in a lifetime now happen more than once a decade. Changes made to the paths of rivers to help prevent lesser floods by allowing water to run more quickly through the system now make it harder to control heavy, sudden flooding. Rather than spreading over a wide area, the flash floods are concentrated by the high flow-rates and therefore produce more destruction where they break through defences.

What no one can doubt is that storm power is on the increase. Although it isn't certain that the frequency of storms is going up, on average storms are more powerful than they used to be, making it more likely they will wreak havoc as Hurricane Katrina did in New Orleans and much of the surrounding territory at the end of 2005. It is telling that months after Katrina, the richest country in the world had been unable to repair the damage that the storm did in a few hours. The scale of the destruction meant that much of New Orleans was still in need of repair eighteen months after the hurricane struck.

The kind of storm damage produced by Katrina will become more common. Alongside more powerful storms in traditional hurricane territory there will be a higher probability of such dramatic weather in previously temperate regions – Canada, for example, can expect more high-powered hurricanes to make land-fall on its Atlantic coast – which may be coupled with a wider occurrence of unusual life-threatening weather phenomena from tornadoes to thunderstorms.

Our bodies can't cope unaided with the weather that climate change is throwing at us – we need all the help we can get.

Solutions

House-hunting in a climate-changed world

There are many factors we take into consideration when choosing a place to live. Is it affordable? Is the surrounding area the kind of place we like, whether that's urban or rural? What are the schools like? Is there a great view or a lovely garden? What's the building itself like? But climate change is likely to add some further criteria to the selection process. The two key indicators to consider are increase in temperature and sea-level rise.

Acceptable heat

Any location that is already at the hot end of comfortable is liable to be unbearable before too long, especially in extreme years. At the moment it is not uncommon for older residents of countries with temperate climates in the northern hemisphere to congregate in the south. You can see this in Florida, or the south coast of England and the southern departments of France. But with climate change, these currently pleasant environments will go too far up the scale. If you are planning to retire somewhere warm, consider moving to a location that is currently pleasant but less hot than these locations.

A safe water supply

As well as being wary of uncomfortable heat, it's also worth taking a look at where your local water supply comes from before picking a location that relies on a lot of shipped-in water to stay habitable. In the past, whole civilizations have vanished when it becomes too difficult to supply water to a region as rivers dry up. There are regions in the world today where a fragile existence is teetering on the brink of devastating drought.

Take much of Arizona in the US. Here you will see lush lawns and fountains in what is by nature a desert state. The cost of this in water terms is huge. With declining amounts of water available from the Colorado River, which supplies swathes of Arizona through a canal, there will come a time – and it's not far off – when the water will simply stop. Short of the highly expensive route of shipping water in by tanker, the cities of Arizona will begin to die. Phoenix will certainly risk having some ashes to rise from. When looking to buy in a naturally dry region, look beyond the carefully irrigated lawns. Where does the water come from? Is the supply sustainable? Such considerations should come high on your list, because this is no longer a 'some time, never' kind of problem, it's more likely to be a 'next few years' thing, and you could see your property plummeting in value, or may even face having to abandon your home.

Staying above water

Then take a look at the sea-level-rise maps (see pages 23–7). While developed countries will make efforts to protect cities like London and New York, areas in danger from sea-level rise are also likely to be the hardest hit by storm surges – witness the impact of Hurricane Katrina on New Orleans in 2005. It's well worth avoiding anywhere that is in danger of flooding from a 1-metre rise in sea level, unless it is a location like London or New York where you can be fairly certain that governments will act.

On the smaller scale, before buying a house, check out the area's flood history. Take a look on the internet at flood information from government sources and water authorities. Speak to people who live in the same street about any flood history. If the house is high on a hill, you are likely to be safe from everything except flood water flowing through your house down the hill – are there any likely sources? Check out the state of nearby rivers and whether they have flooded in the past. If you are near the sea, look into the state of the sea defences, and again, what history of flooding can you discover? Whatever has happened in the past it is likely to get worse in the future.

A particular danger point to look out for is the end of a stream or river, if the water runs through a narrow gorge in the hills above. This can result in a tiny stream having sudden, shockingly deep floods. A good example was the fate of the English village of Boscastle, on the north of the Cornish peninsula, in 2004. The river Valency running through Boscastle is little more than a stream – often just a few centimetres deep on an ordinary day. But on this particular occasion 20 cm of rain fell in twenty-four hours, causing a sudden flash flood that came pouring down the gorge. At the peak around 180 cubic metres of water were flowing through this narrow channel every second. High-water marks from the flood were found to be up to five metres above the normal level of the water. Over a hundred vehicles were washed away and a hundred homes had to be evacuated. As it happens, no one was hurt, but the risk is there. Though a place like Boscastle is beautiful to visit, it would be wise to think twice before moving there (at least to a house on the level of the river).

⚡ POWER TIP
Can your house float?

Should you really want or need to live somewhere at high risk of flooding, it is possible to use technology to help with the problems. In the Netherlands, the country with the best history in the world of understanding and controlling flooding, they are now building amphibious homes that look like any other house, but will float up to keep safe during a flood.

The first of these amphibious houses, built along the banks of a branch of the river Maas, have a hollow foundation that acts to buoy up the whole building. Five-metre-high steel posts anchor the buildings in place – instead of floating off like boats, they ride up the posts as the water rises, then back down again, to their original position, as the flood subsides. The team responsible for the floating houses is now working on office buildings up to a hundred metres high that are constructed on a foam and concrete float that will enable them to survive floods without damage. The Dutch government is also looking at designs for floating car parks, factories and more: in future there may be whole towns that can survive inundation.

Heat

In the house

Essential to keeping cool is having a basic understanding of how heat gets from one place to another. Cut a refrigerator or an air-conditioner down to its heart and what you are left with is a heat pump. These take heat from one place and transfer it to somewhere else. In the case of a fridge, 'somewhere else' is the back of the box – the heat is pumped out of a radiator. This is why you won't get anywhere trying to cool a room down by opening the door of a fridge when it's switched on – any reduction in temperature at the front is balanced by the heat pumped out at the back. Actually it's worse than that. No machine is 100 per cent efficient. Some of the energy used by your fridge is put into that transfer, but some of it will be wasted by losses in the system. And guess how it's wasted – as heat. A fridge gives out more heat than it produces 'cool.'

It's also worth stressing the difference between an air-conditioner and a fan. Air-conditioners, like fridges, are heat pumps. They take heat from the room and pump it outside the building. The result is that the temperature in the room goes down. Fans have no way to get heat out of the room. Instead, they encourage liquids on your skin (mostly water) to evaporate a little faster. A liquid needs energy to turn to vapour, so the result is a temporary drop in temperature on the surface of your skin. You feel cooler, more comfortable – which is great – but there is no change in the room temperature. (In fact, because of the efficiency thing, a fan will slightly increase the room temperature.)

Although air-conditioning is frowned on in terms of its impact on global warming, it's doubtful that anyone sitting in an air-conditioned office, tut-tutting at your use of the technology, should be given too much credence. In temperate countries, where air-conditioning has historically been reckoned an unnecessary luxury, it is time to consider whether or not some of your rooms should be air-conditioned. In a new house, for maximum efficiency, this can be built in. In an existing building you may have to use smaller or portable units. As a compromise to avoid too much impact on global warming, consider using air-conditioning only where absolutely necessary. Try out first the techniques listed on page 50 for dealing with heat in your house without power and see where air-con is still unavoidable.

If using portable units, it is essential to make sure that the warm air that comes out of the venting tube is clearing the building. I have seen air-conditioning units with the hose still in the room, merrily pumping warm air from one side of the room to the other. The hose must go outside, and should be positioned so that hot air streaming from the hose doesn't flow back into the house through an open window.

At the very least, look at bringing your beds to the coolest rooms in the house, even if it means turning the place upside down. The inability to sleep well due to hot nights is one of the most dangerous impacts of a heatwave.

Out of doors

When heatwaves have devastated temperate countries, killing many thousands, it is often because the people are not used to such temperatures and don't take the simple precautions essential to surviving in the heat.

Where possible, avoid going out at all in the hottest parts of the day. If you make use of the measures on page 50 it should be cooler inside than out. If you have to go outside, stick to the shade as much as you can. Don't sit out in the sun. Perhaps the most important thing is to avoid dehydration, which makes the impact of the heat greater, and is particularly dangerous for the elderly and the young. Drink plenty of water, but steadily in small quantities, rather than litres at a time, to avoid water poisoning.

Sun and skin

In many temperate countries, the arrival of hot weather puts out a public announcement – 'time to get out there and strip off.' Sun worship may be pretty rare as a religion these days, but it's still very popular as recreation. If we can't get enough sun at home, we like to jet off to find a sunny beach (increasing global warming as we go) to top up the tan. Unfortunately for us, this attitude to the Sun, regarding sunbathing as healthy and tanned skin as a positive asset, has been recognized for some time to be a flawed and dangerous one.

It should be no surprise by now that the Sun's rays have

the potential to be harmful. Not only does the Sun help you dehydrate, it ages your skin, can cause painful burning, and overexposure to the Sun's rays significantly increases the risk of skin cancer. Australia provides a sobering example for the temperate regions of the world. Many Australian citizens are immigrants from Europe, who have moved to a region with two to three times (Southern Australia) or even four times (Northern Australia) the level of ultraviolet in the sunlight. Skin tones that are balanced for European levels of light fail to give sufficient protection, and the result has been an unusually high level of skin cancer.

With climate change, the amount of ultraviolet exposure received by those living in latitudes above 37 degrees north (which takes in most of Europe, Canada and the US north of Texas) is set to increase. More care will be necessary to avoid risk from the Sun. But strangely, the answer is not to stay indoors or shaded all day and every day. If you feel the urge to buy your children all-over sun-protection suits – resist. Sunlight isn't a universal monster that's out to destroy us. Properly controlled exposure to the Sun is essential to keep up your levels of vitamin D, a vitamin that is important for reducing the risk of osteoporosis (because the body needs vitamin D to absorb calcium), rickets, prostate cancer and breast cancer, and moreover probably decreases the risk of diabetes and multiple sclerosis.

To get an appropriate level of sunlight to keep your vitamin D topped up you need to be outside, not receiving the sunlight through glass. Sunscreen also drastically reduces the vitamin D-producing benefits of sunlight. Sunscreen with just protection factor 8 cuts vitamin D production by over 97 per cent. Another problem is that darker skin pigmentation also reduces vitamin D production. A dark African skin shade reduces production by between 80 and 98 per cent from that in fair Celtic skin. There is also some reduction with obesity and with age (production is reduced to one quarter by the time you reach 70) – in each of these cases, increased exposure to the Sun is necessary to get appropriate vitamin D levels.

A small, controlled amount of exposure to summer sunshine without sunscreen is therefore a good idea, for vitamin D production. The exact approach to take depends on your location. In the latitudes above 37 degrees north, recommendations for those with fair skins are to have exposure around the middle of the day, starting with around two minutes per side and working up to a total of around fifteen minutes a day, exposing up to half the body. These amounts need to be varied depending on your skin type and location. What is essential is that you should cover up or get in the shade if you begin to feel hot or uncomfortable, or if your skin gets a slight pinkness, and that outside this window of vitamin D sunbathing, you should stay in the shade or use sunscreen of SPF 15 or above. It's also best, apart from this exposure, to stay out of the sun in the four hours around the middle of the day. In hotter climates (and in Northern Europe as climate change advances), move the period of controlled exposure away from the middle of the day. For children, or if in any doubt, take medical advice.

As a rough rule of thumb, it should be enough to expose around a quarter of your skin area (for instance your face, hands and arms) for between a quarter and a half the time it takes your skin to take on a pink tinge. In high latitudes in midsummer midday sun, with pale Celtic skin, this amounts to around ten minutes' exposure (ranging up to forty minutes with the darkest skin tones). For the stronger sunlight typical of the southern US states and southern Europe (which with climate change may become more typical of northern Europe) these should be more like five minutes and twenty-five minutes respectively. It's best to get a clear picture of your own circumstance based on around one third of the time it takes your skin to start going pink. Of course, it needs to be emphasized that this process is a matter of balance. You do need some exposure to the Sun, but you don't want to burn.

Sunbathing will carry through to a degree into winter, but it is also worth considering supplements if ultraviolet levels are low. For inhabitants of countries above latitude 37 degrees north, it is impractical to get enough sunlight between mid October and mid April, and ideally dietary supplements should be used. But getting vitamin D from food is not easy. Very few foods contain it naturally. A portion of oily fish (salmon, mackerel or herring, for instance) contains around one half of the recommended daily amount. Milk and margarine are fortified with vitamin D in Canada and the US, but not sufficiently to cover the lack caused by reduced sunlight – and such milk is banned in Europe, as during the 1950s some young children suffered from vitamin D intoxication – poisoning due to consuming excess vitamin D – because of fortified milk. Two teaspoons of cod-liver oil a day will provide about half your vitamin D requirement. Boosting calcium in the diet also helps with vitamin D absorption, though surprisingly wholemeal bread is bad for you in this instance, as it prevents calcium being absorbed – consider eating wheatgerm bread plus more fibre from vegetables and fruit.

POWER TIP

Avoiding wrinkles

One drawback of actively seeking the Sun, even in controlled doses, is that exposure to direct sunlight does increase facial wrinkling. Although it's probably simplest to get your dose of UV via the face, hands and arms, if you would like to keep wrinkling down, then you could keep your face shaded and expose more skin elsewhere to compensate.

One final item to note is the use of antioxidants to avoid the harmful effects of the Sun. These miracle chemicals that seem to help reduce risk of heart disease are also valuable for reducing the impact of ultraviolet. Look for sunscreens containing antioxidants, which show a real superiority in cutting out UV. There is an uncertain benefit from eating antioxidants: some research studies suggest food rich in vitamins C and E can help, others that there is no benefit; but there is strong evidence that lycopene, the chemical that makes tomatoes (and watermelon) red, reduces the damage caused by ultraviolet, so boost the amount of tomatoes in your diet, especially in the summer. There may also be some benefit from superfruits like pomegranate and blueberry, but the jury is out on these.

In the car

Unusually hot weather poses specific threats to the driver. Cars themselves are susceptible to problems, particularly when stuck in traffic jams. There comes a point where the cooling system is struggling to keep the engine at a safe temperature. First aid for overheating, where there is no sign of steam from a leaking radiator, is to turn the car heater up on full. This is no fun for you, but can have a significant cooling effect on the engine. As soon as is practical, get off the road and give the engine a chance to cool.

Overheating isn't good for those inside the car, either. The driver is likely to make more mistakes, whether from the glare of the Sun or just from feeling hot and bothered. Remember how hot the inside of a car gets in direct sunshine with the windows closed. It's a little greenhouse.

Never leave children or pets in cars in full sunshine for more than a couple of minutes (and then make sure there's a safely open window). Use shades to cut down on the amount of direct sunlight coming into the car when parked (even better, try to park in the shade). All metal surfaces exposed to the Sun can give a painful burn when touched – take care.

Drought

Living with drought

Drought is liable to become a fact of life for many more of us as much of the world receives less summertime water. If drought is new to your part of the world, be prepared to take a different attitude to water. Conservation doesn't just mean a hosepipe ban, but using water sparingly and carefully. You may be unable to drink tap water for a period of time when reservoirs are particularly low and liable to pollution.

When water is in extreme short supply, drinking is the only sensible use for that water. Don't flush the toilet. See page 100 for suggestions on building your own toilets, or get a chemical camping toilet to see you through. Reuse water and collect any rainwater that does fall.

Cars, too, can suffer in extreme drought. It's best to carry a supply of water to top up the radiator and to use (sparingly) to clean dust from the windscreen. Don't use the windscreen washers – they are too profligate – use a damp cloth.

Wildfire prevention

Shortage of water isn't the only problem with drought, it's a condition where wildfires are easily started, and can then spread to cover many square kilometres. Take simple precautions to avoid starting fires that could kill and destroy a huge amount of property. Never throw a cigarette or lighted match on to the ground – ensure they are stubbed out and put in a metal container. If you make a fire outdoors, dig a clear area of ground around your fire and sink it a little into the earth to minimize the chance of the fire spreading. When you leave the fire ensure that it is fully put out – cover it with earth to avoid wasting water on it – and never let it get too big.

If you are in a drought area, don't put anything on the fire that will send sparks floating into the air – for instance paper, which can partly light then float away, or wood from resinous evergreen trees that can send showers of sparks out as it flares up.

⚡ POWER TIP
Get it under control

If your fire escapes despite your efforts, or you see an outdoor fire breaking out, take action very quickly. In a surprisingly short time, the wildfire will be beyond your ability to control it.

❏ Smother the flames – if it is still a small fire, use soil, a damp blanket, a sheet of wood, a well-leafed branch or other items that will be slow to catch light to cut off the oxygen and kill the fire before it can take hold.

❏ Use the equipment provided – if you are in a forested area fire-suppression equipment may be provided. Most common are long sticks with a flat rubber blade on the end, or a sort of extended version of a broom. Don't flap at the fire repeatedly with these – the effect will be to fan the flames. Instead, use a measured squashing action, holding the end in place for a second or two, to smother the fire.

Escaping fires

If it is not controlled early on, a wildfire will get out of control very quickly. Once it really catches, nothing you can do as an individual will stop it. If you are in a safe position, ring the emergency services. If there is any doubt about your safety, get away first, then ring.

Unless the fire is right on top of you, take a moment to get your bearings. Even if the flames make it very hot, don't take clothes off – they will give brief protection from any flame that reaches you and will cut down on the amount of heat radiation that hits you directly.

Check which way the wind is blowing (if it's not obvious from the direction that the smoke is being blown, drop a small piece of grass from head height and see which way it floats). Then head away from the flames into the wind (with the wind in your face), as you will be heading in the opposite direction to the way the flames are moving most quickly.

Try to put something between you and the fire that will stop or at least slow down the flames. Rivers, wetland, buildings, barren land, road, fire breaks – anything without dry plant life to pass the flames along with the speed of wildfire. Don't go for high ground – updraughts tend to make fire travel faster uphill than down.

If all your routes are already blocked by fire, look for a region where the flames are relatively low and in a narrow band. If you have any water, douse your hair, exposed flesh and clothes. Try to cover as much of your body as possible with wet cloth, including your nose and mouth. When you have found the best route, take a deep breath and run. If your clothing catches light, roll on the ground to put it out as soon as you are well clear. It may be, however, that the flames are just too bad to get through. If so, find the barest site within the ring of flames, and dig yourself a trench. You will need to cover yourself with earth and hope you can survive as the fire passes over. Bearing in mind this may be the only alternative, check again whether you can't get through the flames.

⚡ POWER TIP
Stay in the car

If you are in a car, stay in it. Fuel tanks are much less enthusiastic about exploding than we are given to expect by the movies. A car will shield you from the worst of the heat. Obviously you should drive away if possible, but if your car is surrounded by fire, you have a better chance of surviving in it than outside it. If the car is engulfed in flames, yes the fuel tank may explode, but often it won't. People have lived through fires that have been so hot that the glass in the car windows began to melt, but only because they didn't get out of the vehicle until the fire had died down.

Cold

In the house

All the essentials have already been covered in the section on dealing with cold in a power cut on page 43. If you aren't in a power cut, you can make use of your heating, of course, but all the advice about insulation etc. is still just as valuable.

Outside

Those who aren't used to really cold weather need to take extra precautions when it begins to strike. Take a lesson from the inhabitants of colder climates. Use layers of clothing to keep warm, and make sure it's breathable material, so that you don't get too sweaty on the inside. Don't go out without gloves and a hat, even though (see page 45) the hat isn't quite as important as some of us think.

⚡ POWER TIP
Avoid touching metal in cold conditions

In cold conditions, metal can give you a 'burn' or stick to your skin. Avoid contact with metal in these circumstances. It's particularly easy to forget this if your car gets stuck. Make sure you are wearing gloves.

Expect the hazards of cold weather. Keep a shovel handy to shift the snow. Try out different shoes and boots on ice – it's not always the most obvious sole that has the best grip. When walking on ice, take short steps and lower your centre of gravity. Be prepared, if you slip, to drop on to your bottom if you can. The result of walking like this is to cause amusement to those around you, but it reduces the chances of a dangerous fall.

After a freeze – in the house

If you are hit by cold weather, particularly if your heating is not operating, there is a significant danger that there will be a burst in the water system. When water in pipes freezes, the ice takes up more space and can break through practically anything. All of your water system should be insulated to minimize the chance of freezing. If you have a water tank in the attic, make sure that there is no lagging between the ceiling below and the base of the tank, so that the warmth from below helps stop the water freezing.

⚡ POWER TIP
Why does ice expand?

It might seem strange that ice takes up more room than water, when we've already seen (page 21) that water expands as it gets warmer, and so should get more compact as it gets colder. In fact water does contract as it cools until around 4°C, at which point it starts to expand again. Water is unusual in taking up more room as a solid than as a liquid (though not unique – acetic acid and silicon, for instance, both have the same property). It does this because when it forms into crystals, the bonds between the atoms are forced to fit in certain positions by the 'hydrogen bonds' that make water molecules act like little magnets. It's a bit like the way you can keep an unmade construction kit in a smaller bag than the volume of a model you make out of the bits.

Make sure that water is flowing in any pumped systems (heating, for example) – a frozen blockage will overload the pump. If you do have frozen pipes, don't try to warm them up with a flame – this could damage joints in metal pipes or melt plastic ones. A more gentle heating approach such as a hair dryer is fine, but make sure there is no chance of a leak spurting water into an electrical device.

Snow and ice – in the car

Have the essentials for keeping a car going in the cold (anti-freeze, ice-scraper, de-icing spray, screen wash) to hand before the really bad weather hits. Allow time to deal with de-icing before moving off. Don't try to drive with just a porthole of cleared glass. Accidents are caused by rushing this process. While working on the car, wear gloves – in very cold weather your skin can stick to the metal and rip off. If you drive a car with a diesel engine, remember that diesel fuel becomes waxy if the temperatures drop low enough. Keep the car somewhere sheltered. If low temperatures become a regular occurrence, consider getting a diesel with an electric engine heater, which can be used to warm the engine before starting.

Similarly, if you live in a region that was temperate but is now experiencing harsher winters, take the lessons of countries where this has been the case for many years. Be prepared to invest in snow tyres or chains, for example.

⚡ POWER TIP

Expect to get out of the car

When going out in your car in snowy conditions, you should assume that you are going to get stuck. Will you survive? Make sure you have enough warm clothing to be OK outside the car. Whether you get stuck in snow or break down on a motorway and have to wait on the verge, you may have to stand up to the weather, so even if you prefer to drive in shirtsleeves, make sure you have plenty of warm and waterproof clothes (and stout shoes or rubber boots) with you. Keep a shovel and a bag of sand or salt/grit mix in the boot, along with usual breakdown essentials like a torch, first-aid kit, high-visibility jacket and warning triangle.

If the car gets stuck in snow – and even a thin layer can make it surprisingly difficult to get started, especially up a slope – clear the snow as much as possible from in front of the wheels, particularly the driving wheels. See if you can lay something in front of the powered wheels on the car to give them traction. Apart from sand or grit, you could use a sheet of cardboard or break some twigs from surrounding bushes. If all else fails, an item of clothing might help.

Don't go heavy on the accelerator; this will just make things worse. Try to start off in second, or even third, gear if your car is capable of it. In deep snow, you might be able to get some purchase by rocking the car back and forward between forward and reverse, though it's best to clear a track in front of the wheels.

If you are immovably stuck, run the car engine for up to ten minutes in each hour to pump heat into the car. (Make sure the snow is well clear of the exhaust.) Minimize the drain on the battery – don't run the radio or interior lights when the engine is off (and don't use the headlights or heated rear window at all). Try not to go to sleep – there is increased risk of hypothermia as your body temperature drops – and keep away from alcohol (see page 46).

Once you do get moving, use extreme caution on icy roads, particularly if you aren't experienced in cold-weather driving. Keep in as high a gear as possible. Don't accelerate and brake more than you have to – and remember that when you do stop, it could take you up to ten times the usual distance on the ice. Over-anticipate. If you are approaching a T-junction, for instance, assume that your brakes won't stop you. Slow down as much as possible without heavy brake use.

However careful you are, at some time you are likely to skid. Be prepared – there isn't time to think about this when it happens. Resist the urge to brake hard. If the front wheels go (usually from accelerating in a bend, or turning the steering wheel too sharply), don't brake. Steer into the skid with a rear-wheel-drive car, but with a front-wheel drive keep the wheels in the direction you want to go and a slight contact on the accelerator. As soon as you regain control, straighten and carefully accelerate. If the rear wheels go (usually going too fast on a bend), don't brake or accelerate. Turn the wheel gently in the direction the rear end is going. When you are back in control, steer normally and carefully accelerate.

If you brake too hard you might start to spin. Lighten up on the brake pedal until the wheels start to turn and you can regain control of the steering. When you've recovered, consider buying a car with anti-lock brakes.

Storm

In the house – preparing for floods

If, despite all the doubts, you fall in love with that cottage by the sea in a surge danger area, or a desirable residence built on the flood plain of a river, at the very least you should make some basic preparations for flooding. If you live in such an area, keep your defences ready. Ensure that you have enough sandbags to deal with main sites where water can get in. In many countries these are provided by local government – otherwise you can buy them or make up your own. In an emergency, carrier bags filled with earth will provide a (literal) stopgap.

Make a survey of the outside of your house. Where could the water get in? Note any vents, airbricks, flues, sink outlets, overflows and other links from the outside to the inside. Remember easily opened flaps like the letterbox and any cat or dog flaps. Have simple means of sealing up these openings ready for use in an accessible place. But don't seal them up when there is no imminent flood. Blocking up an airbrick can result in damp encroaching in your house or risk danger from poor ventilation of heating equipment. Blocking a heating flue has equally obvious dangers, and should only be done if you are sure the heating is switched off and will remain so until it is unblocked.

In the house – living through floods

The best, if rather simplistic, recommendation is not to live in a flood risk area (see page 138) – but for many people this is not an option. If so, it's essential to have a plan for the action needed when you get a flood warning. See page 165.

If the water is rising very quickly and you don't have time to get clear before it's more than knee-deep, or if there is no high ground nearby, so getting out of the building won't help you, don't run for it. You are safer inside a well-constructed building than out of it. After taking the precautions on page 153 to block entries and turning off the power and gas, move higher up the building. Make sure you have as much as you can with you that will attract attention: something to make a loud noise (an air horn, whistle, or smoke detector, which can be used as a noise-maker by pressing the test button), a torch, flares (in the unlikely event you have them), a mirror to reflect the sunlight and bright-coloured cloth to wave.

Don't go higher than you have to – take it a floor at a time. If you are forced to move on to the roof, take extreme care if it's not flat. Use ropes or knotted blankets to make safety lines to fix those taking shelter to a robust anchor point like a chimney stack. Always keep a lookout for rescuers and be prepared to signal for assistance.

⚡ POWER TIP
Be wary of water

After the flood, don't assume the mains water is safe – check with the authorities as it may have become polluted. (Certainly don't try to drink the flood water: treat it as sewage.) Try to keep out of the flood water as much as possible to avoid contamination. As the waters subside, watch out for hidden debris and don't switch the power back on until you've been told it's safe – get electricity cables and gas pipes checked for damage.

Flash floods and tsunamis are less predictable – you may get little or no warning. Don't hang around to see if things are going to be OK: get out, to high ground and away from the route of any flood waters. Even a small flash flood can sweep cars away. Get out of a stalled vehicle, don't hang around to go with it. Fast-moving water is very powerful and it does not have to be very deep to carry you away. Water well below your waist in depth can be enough to blast you off your feet. Just don't risk crossing a flood stream unless it's life or death. If it really is necessary, take the advice below.

Watch for any gullies that could carry sudden floods. Even if you are well above sea level, you could still be hit by heavy runs of water from higher ground. Be wary of walking anywhere in the flood water, even if it is calm. Not only could there be another surge, but the water is likely to carry foul pollution, and also you won't be able to see if it has washed away the surface beneath – you may suddenly plunge into an unexpected abyss.

In the car

One hallmark of the climate change we are facing is heavier downpours, and this makes flooding a likely occurrence on many roads. Some can accumulate a surprisingly deep flood pool in minutes. Drive as slowly as possible through any pooled water – watch other cars to get a feel for depth (if there is no other car on the road, take it very carefully). Where you are faced with a deep pool, drive along the middle of the road – this is usually a little higher than the sides, to help water run off. Keep in a low gear and don't change gear as you go through. If there's spray up the side of the car you are going too quickly.

Go too fast through a flood pool and you could skid or, worse, get water up to the engine and stall or cause expensive damage. You will also look very stupid. As you come out of the pool, try your brakes. If the water reached them, they may not be working very well. Put repeated gentle pressure on the brake pedal until they are back to normal.

Crossing fast-moving water

The simple solution is not to do it. A deceptively shallow fast-moving stream of water can sweep you off your feet. If, and only if, you absolutely have to get across, take these precautions:

❏ Look first – check for obvious obstructions. Try to find wider sections of the stream, which are likely to be shallower and slower moving. Watch out for obstructions under the water that might make your crossing dangerous. Cross upstream of an obstruction – that way you can use it for support and you avoid the dangerous eddies that are likely to form downstream.

❏ Check the water temperature – if it's freezing cold don't get into the water, you wouldn't survive long. Try to find something to float across on.

❏ Get yourself a good-sized stick – use this both to test for the depth of the water and to help keep your balance.

❏ If the water is shallower than your waist height, consider taking off your trousers or skirt. You will be more comfortable on the other side, and you will present less resistance to the water (especially if you have to swim). This isn't a time for modesty.

❏ If you have a long rope and someone on the far side to help, try to get this across the water to help you keep stable.

❏ Find something that will act as a flotation aid if you get swept away. Ideally this should be something you can carry with your hands free. If there's nothing purpose-built, look for thick sheets of polystyrene, large strong plastic bags or airtight clothing you can fill with air and tie off, large plastic bottles and similar makeshift floats.

❏ If there's more than one of you, go in a chain, hands on the hips of the person in front, to reduce the chances of anyone getting washed away.

❏ Once you step into the water, don't try to go quickly. Turn sideways on to the water flow and shuffle your feet along rather than lifting them clear.

If the water is too deep to wade, keep your clothing to a minimum when swimming across. Always use something lighter than water to help you keep above the surface, even if you are a good swimmer. However capable you are, you will be swept a considerable way in the direction the water is flowing, so allow for this in choosing where you go in, both in terms of having somewhere safe to get out, and in making sure you aren't swept into any obstructions.

Avoid death by tourism

 If you hear news of a dramatic storm of any kind – tsunami, tornado, hurricane, thunderstorm – resist the temptation to go and watch. It's not uncommon for onlookers to be killed.

It is practically impossible, unless you are on high cliffs, to be in sight of a tsunami without being at risk. Other storms like tornadoes and thunderstorms are notoriously difficult to safely track. You may find the storm doubles back on you, placing you in trouble before you can do something about it. If the storm warning is for evacuation, evacuate. If it's to take shelter, do just that. It really isn't worth risking your life to see some impressive weather.

This warning doesn't just apply to huge tropical storms. In temperate countries, especially with sea-level rise, it's quite possible for a storm surge to send huge waves breaking over the sea front that will wash away onlookers. Keep safe.

Hurricane warning

While there is no evidence that global warming is increasing the number of hurricanes, it is very likely that it is increasing their power. There are more severe storms out there than there used to be. True hurricanes currently stick to the tropics (though this may not remain the case if climate change continues). They are immensely powerful storms, which can whip wind speed up to 300 kilometres per hour and wreak havoc when making landfall.

There is a certain amount of confusion over just what a hurricane is. It is a powerful storm that arises out at sea, forming huge slowly spinning spirals that are often as much as 30 or 40 kilometres across, and have been known to be 500 kilometres wide. The spin is usually anticlockwise in the northern hemisphere and clockwise in the southern. They can drift for days or even weeks before finally dispersing, usually after making landfall. Although hurricanes are very obvious on weather satellite images, their paths of destruction are hard to predict as they can suddenly veer, or even double back on themselves. Part of the confusion arises from the way that the same phenomenon is given different names in different parts of the world. Though they're hurricanes in the North Atlantic, Caribbean and parts of the Pacific, they are known as cyclones around the Indian Ocean and as typhoons in the rest of the Pacific and the China Sea.

⚡ POWER TIP

Know your shelter

If you live in hurricane territory and don't have your own hurricane shelter, make sure you know where the nearest emergency shelter is and how to get to it in a hurry. Although there is often warning, it may be limited. When visiting the tropics, check where your hotel's shelter is as soon as you arrive.

Keep the surroundings of your home clear of objects that the wind can pick up and hurl at the building – and that's pretty well anything. If the storm manages to get through, use whatever you can to defend yourself from flying debris. Don't go outside if the storm has already arrived, where you could be swept away or hit by flying objects and collapsing walls, and don't try to drive away. A car is not heavy enough to keep you safe. When the wind dies down, give it a little time before assuming the worst is over. The centre or eye of the storm is calm. In many cases you may never experience the eye, and only the edge of the hurricane brushes across you, but if the eye does pass over you, then you have just as much of a storm to come again, only blowing in the opposite direction.

Don't be tempted to go out and take a drive around to look at the damage right after the storm. Assess your own position, but give the emergency services a chance to get established before heading out.

Although it too involves circling winds at high speed (some have been measured at over 600 kilometres an hour), a tornado is a very different prospect to a hurricane, typically extending no more than 50 metres across at the base. As already mentioned, the occurrence of big tornadoes may well not be on the rise, but the everyday sort experienced all over the world may be becoming more common. The impact of a tornado is less widespread than that of a hurricane, but where it does hit the effect is much the same, as are the precautions to take. The main difference is the danger of being sucked up into the tornado, rather than blasted along by the hurricane. If trapped outside, keep as low as possible, try to find a shelter like a cave or gully, and protect your head against airborne missiles. Tornadoes are usually very visible from a good distance. Get out of the way. Move as quickly as you can, away from the direction the storm is moving in (away from the tornado if it's moving away from you; at right angles to the tornado if it's moving towards you). Tornadoes can change direction, but this is still your best bet.

Inside, close all windows and doors on the side facing the

approaching storm, but open them on the opposite side to balance out any build-up of pressure that could cause the structure to collapse. Tornadoes are particularly partial to hurling roof tiles, making lethal slicing weapons of them. Make sure you are well protected by solid objects from any incoming roof sections. If possible head for a storm cellar or shelter. Make sure you are inside a structure that can stand up to the impact of a tornado. This does not include caravans, mobile homes and trailers, which can be ripped apart, or cars, which can be thrown up in the air. For a small tornado, such as those typical in Europe, being in the middle of the ground floor of a conventionally built brick or stone building should be fine. Where there are big tornadoes, ideally the building should be specially built to withstand tornadoes with a steel-framed structure, or high-grade reinforced concrete.

Lightning strike

More powerful storms may also imply more thunderstorms. Lightning strikes are not uncommon, and the sheer power of a bolt of lightning is a force to be reckoned with. If a thunderstorm approaches, get off high and exposed ground.

> If you are the highest thing around, you are the natural target for lightning. Avoid single trees and metal objects (such as masts and pylons) that could act as conductors.

Get to shelter as soon as possible. Do not shelter in the opening of a cave, which can attract a lightning strike. It's fine if you are well into the cave, at least a body-length away from the opening. If you can't get indoors, a car is an excellent shelter – the metal cage it provides prevents the electrical charge from getting inside. Make sure when getting out after the storm, though, that you don't touch the outside of the car and the earth at the same time, in case there is a residual charge.

If you don't have time to get clear, you can't dodge lightning. Some survivors have reported feeling a tingling sensation and their hair standing on end just before a strike, as the potential builds up. If you experience this, drop to the ground and lie flat. Get as low as possible and avoid being in a position with a route through your heart being the natural one to the ground (for example one hand on the floor and one in the air).

You may have heard that it's dangerous to use the phone during a lightning strike. It's not quite the hazard that some make out, but on average one person a year is killed this way, and quite a few more are injured. Wireless home phones shouldn't present a risk if when you use them you are well away from the base station – though that could be fried – and mobile phones are fine. Any electrical equipment that has outside wiring or aerials is at risk of attracting a lightning strike – so it is best to stay away from a wired phone, the TV and the computer in the event of a nearby thunderstorm.

⚡ POWER TIP
How close is the lightning?

The old counting trick to establish how far away a thunderstorm is (and hence the likelihood of a lightning strike) really works. You are particularly at risk if the flash of lightning and the crash of thunder appear pretty well simultaneous. The greater the separation between the two, the further away the storm is. It's all down to sound taking longer to reach you than a flash of light – both happen simultaneously in the lightning bolt.

Light travels at 300,000 kilometres a second, sound at around 340 metres a second. To all intents and purposes, you can think of the light as arriving instantly, so it's just a matter of timing the sound's journey. Each second you wait it will have travelled 340 metres, so 3 seconds means the storm is 1 kilometre away, 6 seconds that it is 2 kilometres away and so on.

Remember also that your electronic equipment that is connected to outside wiring is susceptible to damage from power surges during a lightning storm. Ideally turn off and disconnect your computer, modem (or broadband router), TV, satellite or cable box and other TV equipment like recorders. Disconnect all landline phones except one cheap phone you don't mind being damaged. Note that 'disconnection' should ideally be removing the connection from the aerial, dish or telephone line as well as unplugging from the mains. This disconnection should not be done when the storm is nearby. If you have left it this late, just stay well away from the equipment and hope – you don't want to be unplugging something when there's a strike.

Another risk, low-level but one that has occasionally caused harm to people, is that a huge strike may send a power surge up water pipes from the ground – ideally, then, stay away from sinks, baths and other plumbing when the storm is directly overhead. On the whole, hiding under the duvet in the middle of a room (definitely not on an iron bedstead), reading a book by battery lantern, seems your safest distraction in a storm.

Resources

Air-conditioning

If you live in a temperate area that is undergoing temperature increase, you might in the past have been doubtful about the benefits of air-conditioning. Overcome those doubts. It's particularly difficult if you are environmentally minded, as air-conditioners definitely add to the weight of global warming – but so do pretty well all other essentials of life; it's just in some countries air-conditioning hasn't traditionally been thought of as an essential.

Consider getting properly installed building air-conditioning for at least some rooms of the house – perhaps the lounge and bedrooms – as a minimum. If this isn't practical financially or in the timescale, look at portable air-conditioning units. These are much more efficient than they used to be, provided you make sure that the warm air is venting properly to the outside.

Similarly, air-conditioning in the car should be considered as important as a heater, as climate change pushes up summer temperatures. Although the air-conditioning does increase fuel consumption, it uses less extra fuel than driving with the windows open, while providing a much more pleasant environment. A car can get like an oven in strong sunlight, both causing a risk to health, particularly for the very young and elderly, and making the driver more likely to have an accident.

If you haven't got air-conditioning at home and the temperature is getting oppressive, make use of air-conditioned public buildings, from shopping malls to libraries, to cool off. See page 50 for more on cooling off without using power.

Fans

Remember that fans don't cool a room down; they only reduce your skin temperature by evaporating moisture. This means they are very much second best to an air-conditioning unit, but fans are cheaper, can be put pretty well anywhere, and are better than nothing. As there is no impact on the room, only on your skin, it's essential that air from the fan blows directly on you.

Water

Bearing in mind the importance of having a good water supply when there's a risk of drought, look at opportunities to increase your capability to store water when you are making changes to your house or building a new house. Consider a pond in the garden. Go for a larger water tank. Capture rainwater. See pages 124–5 for more water resources.

Insulation

Winters may get colder as a result of climate change – and with less reliable energy sources to maintain your heat it's essential to make the most of the insulation you have. In the longer term, get more insulation in, but in an immediate emergency raid insulation, if necessary, to ensure that pipes are properly lagged to avoid bursting.

CHECKLISTS

Car cold-weather kit

Keeping your car ready for the cold is doubly important if you might have to rely on it to get to safety. Ensure you are well prepared with the basics:

Anti-freeze .... ❏
Ensure your car's coolant contains the appropriate anti-freeze.

Windscreen wash fluid .... ❏
Keep your washers working and not frozen up – essential given the dirt that gets thrown up from wet roads.

Garage .... ❏
If you can, get your car into a garage to make it instantly driveable with no delay for scraping off ice and snow.

Scraper .... ❏

In an emergency a credit card isn't a bad substitute, but a window scraper does the job more quickly. This is better than a de-icing spray, which cools the glass, resulting in misting up.

Snow shovel .... ❏
Keep a snow shovel in your car in case you need to dig your way out.

Bag of salt and grit .... ❏
Useful to sprinkle on the roadway in a line ahead of your tyres to help you get started.

Warm clothing .... ❏
In the car's warm interior it's easy to get blasé about the outside temperature. But if your car breaks down you may soon get cold, particularly if you break down on a motorway/highway and need

to take shelter on the verge. Always have a warm, waterproof coat, scarf, waterproof footwear, hat and gloves in the car in cold weather in case you have to get out unexpectedly.

Jump leads .... ❏
With the extra load from more use of headlights, heated windows etc., car batteries go flat more often in cold weather. Carry a set of jump leads so you can give or receive help when a battery is dead.

Mobile phone .... ❏
This is not exclusively a car requirement: never go out without a mobile phone in severe cold weather.

Flood action plan

Move as much as possible to upper floors .... ❏
Move as much as you can of your cherished items – assume everything left below will be destroyed.

Get a food store upstairs .... ❏
Make sure you have food and drinking water to last a day or two.

Portable heating and lighting upstairs .... ❏
Assume you are going to have to do without mains power. Get as many battery lights together as you can. If you have a portable heater suitable for indoor use, take that upstairs too.

Pets and their requirements moved upstairs .... ❏
Don't forget your pets in the pressure to get safe.

Sandbags .... ❏
Use sandbags to block any gaps where the rising water can get in. This will definitely include doorways and may be needed for windowsills as well. Sandbags may be available from your local council. Alternatively you can buy or construct them (see page 153).

Turn off electricity and gas .... ❏
Reduce the risk of danger by switching off electricity and gas at the mains before the water gets to them.

Make sure there's plenty of clean water upstairs ...... ❏
Floods often bring with them a horrible polluted sludge of sewage and waste. Make sure you have enough drinking water to cope.

Find somewhere else to go .... ❏
If possible, don't let yourself get trapped upstairs. Make your property as safe as you can, switch off the power, then lock up and get away to high ground. It might be distressing to leave your house to the mercy of the flood, but at least you know that you won't need rescuing.

Hurricane action plan

Listen to local radio/TV for hurricane warnings .... ❏

In the event of a warning being issued:

Consider evacuation if you are near the coast or in an area that is liable to flood .... ❏

Continue listening to radio/TV for further information .... ❏

Cover windows with boards to stop flying objects breaking through .... ❏
Don't go near the windows in a storm, even if they are boarded up.

Make sure your car is filled up with petrol and positioned for a quick getaway .... ❏

**Prepare for storm damage – assume you will
lose water and power** .... ❑
*Store drinking water and be prepared for a blackout (see
Chapter 2).*

Make sure any children and pets are safe .... ❑

Take shelter in the safest part of the house .... ❑
*Apart from staying away from windows, look for potential
hazards – avoid the area where a blown-down chimney or
ripped-off roofing might fall through the house, for example –
and look for structural features like staircases that might help
provide shelter.*

**If outdoors when the storm hits, open
doors carefully** .... ❑
*Not only is there a danger of the door blowing off, there have
been instances where the sudden inrush of wind into the
building has blown out windows or parts of the wall opposite.*

FIVE

SAFE AND WELL

With all the potential impacts of climate change, the risk of having to fend for yourself, and to keep your family safe, is greater than ever.

What to do . . .

... to stay safe

... to prepare for chaos

... to avoid insect attack

... to get out of a building when in danger

... to defend yourself

... when someone is injured or needs assistance

Safety first

Back in the 1960s, when disaster novels like John Wyndham's *The Day of the Triffids* were all the rage, there was one thing everyone knew for certain. If our fragile veneer of civilization were peeled away, we would all have to defend any resources we had against violent attack from other human beings. It would be everyone for themselves in a battle for survival.

The impact of climate change may not be as drastic as that caused by Wyndham's killer plants – and we certainly have more warning that the problem is on the way – yet it is quite reasonable to assume that there will be increases in violent crime and looting, and even that our homes may not be safe havens.

As the risk of a violent break-in is likely to increase in these difficult times, it is essential to know the defensive strengths and weaknesses of where we live, and the best way to protect a location under attack. Despite years of crime prevention advice from the authorities, many houses are poorly protected against break-in. It's time to rethink the security of our homes. It is also important that our families know how to get out of the house when in danger, and know where to meet up if things go wrong.

Historically, unrest and civil disruption have frequently been triggered by blackouts, heatwaves and storm damage, all symptoms of global warming. With this in mind, it's important to be aware of how to stay safe on the streets. Take the most central aspect of climate change, increased temperature. The blue skies of a sunny day make us feel more positive, but once the heat kicks in, that pleasant feeling of well-being is easily pushed aside by aggression. Prolonged heat, particularly in the city, can provoke irritation followed by violence.

Studies at the University of Glamorgan have shown that there is a link between hot weather and increased serotonin levels in the brain. Upping the quantities of this neurotransmitter can result in more aggressive behaviour. In the hot summer of 1988, for instance, the murder rate in New York shot up by 75 per cent. In experiments, subjects will administer painful electric shocks to others with much less concern if the temperature is high, while riots seem more likely to break out in hot weather. Cities can already be dangerous places where it helps to be streetwise. With the impact of climate change, the streets and even our office blocks could become still more threatening. It's best to know how to stay away from trouble – and how to get out of it if you end up in the thick of a riot.

Danger can strike when you least expect it. Your office block is under attack – what action can you take? Your car is blocked as you take some precious food back home – how will you get through?

First-aid training could make the difference between life and death.

Being able to handle yourself in a crisis becomes much more relevant as climate change bites. If you get a chance to take self-defence classes, so much the better – there is a limit to how much can be learned without practical tuition. And there is another type of training that is well worth investigating. Training that could make the difference between life and death. That's first aid.

Medical considerations come into this chapter because security implies more than simple physical protection. How would you manage in a medical emergency without access to a hospital or doctor? Global warming will put more stress on already stretched health services, while electrical blackouts and other results of climate change will make it harder to get medical help from the professionals. You may have to go it alone for some period of time. The ideal is for everyone to have first-aid training and to be aware of what action to take if someone is injured, whether in an accident or an attack. This book can't substitute for that training, or a detailed first-aid manual, but it can give essential guidance on dealing with medical emergencies.

Solutions

City street survival

The usual rules for keeping safe on the city streets become even more important under the potential pressures of unrest sparked off by climate change. When out in the city, don't provoke others (even if the provocation seems totally unimportant to you). Specifically:

❐ Avoid eye contact – this can be taken as a provocation.

❐ Assume everyone in sight on the street is a potential enemy. This sounds extreme, but the alternative to being on your guard is being at risk.

❐ Don't display team colours: – This could be literal – for example wearing the colours of a football team that has local rivals – or metaphorical.

❐ Don't display conspicuous consumption – in stressful times there is likely to be more resentment of the blatantly rich. Keep your electronics and jewellery out of sight.

❐ Don't appear weak and vulnerable. Walk upright and with confidence. Look ahead, not at the ground. Never seem uncertain about where you are, or where you are going.

❐ Don't comment on passers-by – your remarks could be overheard.

❐ Keep to well-lit and populous streets – a shortcut down a back alley or along a canal bank could be a quick route to trouble.

❐ Consider carrying a dummy wallet, containing a small amount of money and out-of-date credit cards. That way, if you are mugged, you won't lose the important stuff.

❏ Don't use the obvious pockets – keep your wallet in a front trouser pocket, under several layers of clothes or in a money belt. Don't put anything valuable in your outer pockets or back trouser pockets.

❏ Don't get in an unknown car, even if it claims to be a taxi (only use clearly licensed taxis). If someone stops a car and calls to you for help, speak to them from a safe distance in case they intend to pull you into the car.

In touch with your surroundings

Be aware of your surroundings, particularly if there is tension in the air, or disruption of normal law and order because of the impact of climate change. The most dangerous time for street attacks is in summer, especially during a heatwave – a situation that is likely to get more common in many countries – and during the evening and night. As you walk along, scan around you. Keep your eyes moving; check for potential trouble.

Though avoiding eye contact is best if you aren't addressed, look confident, not like the sort of person in a classroom who always gets picked to answer the question because they are staring so hard at the floor. This is particularly important if someone speaks to you: then make strong eye contact – don't keep looking at the ground, look them in the eye. Speak clearly, coolly and confidently; don't mumble.

Watch out for threatening-looking individuals or groups and if possible change route to avoid them. Keep to busy, well-lit streets. Of course most people out there aren't a danger – but some will be. It seems harsh, but your first loyalty in such circumstances has to be to yourself and your family and friends (if you are accompanied). This is one time when adopting 'city indifference' and not getting involved is a good survival trait. If you see someone in trouble, get help, call the police – but don't try to be a hero and get involved yourself.

> **POWER TIP**
> ### Distinguish between
> ### fear and panic
>
> In a dangerous situation, fear is a good thing. It's sensible to be aware of the dangers you face, to assess the risk and to be cautious. But distinguish this from panic. When panic takes over, you lose rational control and your chances of survival plummet. Don't criticize yourself (or others) for feeling fear – use your fear to maximize safety – but do try to control panic.

Preparing for chaos

When everything has broken down it's too late to prepare. Make sure you think through your survival before everything goes wrong. Everyday matters become a lot more complicated without power, communications and transport.

Have a very simple disaster plan that is known and understood by all your family. The default action if things start to go wrong is to meet up at home, but also arrange a meeting point that is sheltered and hidden away, in case things go wrong and you can't meet up at your house for any reason. That way, if communication channels aren't available you can still get together.

Make sure that all your family members who are old enough to be involved know the basics, from the location of the first-aid kit to how to switch the electricity and gas supplies off (and on). If at all possible, get as many of your family as you can through a simple first-aid course too.

Go through the checklists at the end of this and the other chapters. In particular make sure that you have an up-to-date first-aid kit and a kit that you can grab and run with if you need to evacuate.

The driver in a crisis

If the situation deteriorates when you are in your car, remember that you do at least have a seriously powerful weapon and defence in the car itself. Make sure all the doors are locked, particularly if you stop at traffic lights or for an obstruction. When you get out of the car, try to do so in a well-lit environment. It may seem unfriendly, but don't stop and offer help to other drivers who appear to be having trouble – signal to them that you will call for help (and do it), but don't stop.

If there is an obstruction you have to stop for (including a person in the road), don't get out; lock the doors and close the windows. Keep the engine running and the car in gear. Be prepared, if necessary, to reverse out of the situation. As soon as you can, get around the obstacle and away – then call for help.

⚡ POWER TIP
GPS to the rescue

Should you need to drive through a dangerous environment – and that could be anywhere urban if law and order break down – make sure you know where you are going. Stopping to look at your map in a high-risk area is just asking for trouble. It's highly recommended you get yourself a GPS (satellite navigation) system. That way you won't have to be struggling to read a map as you drive, and should your route be blocked and you deviate from it, the system will find you an alternative.

If you haven't got sat nav, make sure you really familiarize yourself with the route before taking it. Look on a detailed map for buildings you can recognize to help keep yourself on track – but this is very much second best.

Provided you act quickly enough, your car gives you a huge advantage over individuals on the street through its sheer weight and power. You are most at risk from projectiles being thrown or shot through the glass, attacks on the glass with clubs (baseball bats, for example), fire bombs and the car being tipped over. Before this last form of attack can take place, a group of people has to be able to assemble along the side of the car and lift. If your engine is running and you are determined, you can get away before this is possible – unless the car has been blocked in by heavy objects that are impossible to move with the power of the car alone.

Keep the engine running. Lean on the horn to warn those around you that you intend to move – you don't want to give them time to attack, but equally you don't want to mow them down. While still keeping the horn blaring, begin to move. Start relatively slowly. However, if individuals move towards the car instead of away, or look in danger of throwing things, then you come to the point where your survival is more important than the safety of bystanders. Speed up to avoid giving bystanders the chance to grab on to the car. Get out of there at the maximum speed at which you can still keep good control – you will just make things worse if you try something dramatic (like a movie-chase-style high-speed reverse) which then results in a crash that immobilizes the car and leaves you at the mercy of an angry crowd.

If you've company in the car, get them to watch out for missiles being thrown at the vehicle. One possibility is being hit by petrol bomb. Don't panic – real cars don't burst into flames as quickly as they do in the movies. Try to put the flames out by sharp acceleration. Abandon the car only if it seems that the flames have taken hold and aren't going to blow out.

When faced with a blockage you can't get through, stop and barter. Offer cash, and if that fails, your possessions. Be prepared to run for it if it seems likely that this will be necessary.

Hitch-hiking in a dangerous world

 Hitch-hiking is dangerous at the best of times, more so when the rule of law is threatened. One simple rule each for hitch-hikers and drivers who are thinking of picking up a hitch-hiker.

1. *Don't hitch-hike.*
2. *Don't pick up hitch-hikers.*

It's not worth the risk from either side. Even if the driver looks harmless, or the person you consider picking up looks as if (s)he wouldn't hurt a fly (even more so if he or she is very attractive), don't do it. Apart from anything else, the person you see might not be the only one involved.

Personal networks

The biggest mistake most people make when attempting to survive is trying to do it alone. Make use of your personal network of friends and relations. Even acquaintances can be valuable. Don't feel guilty about asking other people you know for help – there is opportunity for mutual benefit. When the rule of law is in danger of breaking down, big groups are safer than individuals. Stick together. Try to increase the number of people living in your home. Take a group of people to undertake a task if there is any chance of being attacked. Help each other to your mutual benefit.

The benefit of a personal network isn't just one of extra bodies in self-defence, however. Make use of each other's expertise. (See page 119 for the particular advantages of adding a farmer to your personal network.) If there's someone in your household with a useful skill, use that skill to help others in the hope that there will be some returned benefit. Whatever the circumstances, a good personal network can be a huge advantage.

Securing your home

No one wants to turn their home into a fortress, but should there be a state of increased tension as a result of the impact of climate change, it is only sensible to ensure that your home is as secure as possible.

Take a quick survey of your house or apartment, looking for ways in. Does each door or window have a secure lock? It's surprising how small a window can be used to get into a house – make sure upstairs windows and smaller windows still have locks, and that they are used. Check for other ways in from old coal chutes to dog flaps. However good your locks, though, they won't prevent a window being smashed. Consider metal shutters – these needn't be unattractive, and can be retracting (though if they are electrically powered, make sure there is a way to operate them manually in case of a blackout).

Where there is glass that can't be protected with shutters – for instance, in a conservatory – make sure there is a 'bulkhead' between that room and the rest of the house. The internal door should be high quality and have an outdoor standard lock so that even if someone gets into the conservatory they can get no further.

Patio doors should be bolted at both ends to avoid being levered off from the 'wrong' end. Ordinary doors should not rely on a single lock bolt, as this becomes a weak point that can be forced. UPVC doors usually come fitted with a multi-bolted locking system. On conventional doors, add bolts and/or rim locks.

Look for locks that deadlock. This means the bolt can't be forced back in from the outside. This is especially important for Yale-style locks that normally slip back to allow the door to close. Unless these are deadlocked, they are little use at preventing a determined person from getting in. Don't rely on a rim lock (the sort that fixes on to the side of the door) alone – always have a mortise lock (the sort that is built into the door itself) for extra security.

Remember that the weakest part of a door may well be

the frame. If necessary reinforce the frame. You might also consider hinge bolts, also called dog bolts, which stop the door being forced open on the hinge side. Some UPVC doors come with these as standard, but most ordinary external doors won't have them.

If your house, like many US houses, is built of light-weight material that can be cut through using a power saw, consider moving to a house made of stone, brick or concrete. This may seem extreme, but there really is no way to provide much security for a house made of thin wooden boards (thick logs are OK) or fibreboard panels that can be penetrated with a cheap chainsaw.

Make sure your garage and outbuildings are locked, not just to prevent things from being stolen, but also to make sure they don't provide an Aladdin's cave of breaking-and-entering tools from ladders to garden spades (which can be used to lever open patio doors).

⚡ POWER TIP

Intruder lights

Outdoor intruder lights will deter some who want to keep a low profile, though one of the significant problems of the breakdown of the rule of law is that it becomes less important not to be seen. Even in ordinary times, I did once witness the bizarre sight of a thief attempting to steal my neighbour's security lights: they hardly proved a deterrent then. Even so, outdoor lights switched on by motion sensors can alert you to intruders and make them feel less inclined to hang around.

Strange callers

If the security risk is high after the impact of climate change, be wary of unknown callers at your door. Don't open the door to them, even if you have a safety chain – a well-placed kick can pull that off and open the door. Fit a peephole and address the unknown people through the door. If, for example, they claim that they need help because of an accident, offer to call the ambulance or police for them without letting them in. If they argue, saying they need access to something from your house, the more they argue, the more suspicious you should be. Stick to the point, and don't get into a conversation with them. Call to someone else in the house (even if there isn't anyone else home) to make it seem as if you have reinforcements.

⚡ POWER TIP

The invisible back-up

Sometimes someone who doesn't exist can be a big help. The invisible helper might be a recording of a dog barking, lights and a radio on a timer that switches on when you're out, or just a person you appear to call to when there's someone suspicious at the door. Your invisible helper may not entirely do the trick, but the apparent information that there are other people in the house can make you seem less vulnerable.

Under attack at home

It's hopefully unlikely, but it is possible that your home will come under attack if the changes brought about by global warming result in reduced enforcement of the law. In the event of attack, get people together. If you significantly outnumber the attackers and they don't have firearms, you may want to make your numbers obvious. Otherwise, keep as much under cover as possible. Use defensive measures, such as pouring boiling water or oil from upper-floor windows if the attackers don't have projectile weapons.

Stay in the building while you can. You are on familiar territory, and have the protection of the building around you. Check the external doors and windows – if necessary reinforce them with a barricade. Always be aware, however, that you may need to get out – so keep a line of exit, furthest away from the attackers, clear. If you need to get out, see the following item.

Under siege

If law and order breaks down, whether it's due to a blackout or other problems and shortages caused by climate change, you may find yourself trapped in a building with unfriendly individuals outside. Look for opportunities to keep safe within the building – it may be easier to stay secure inside, especially if you are on familiar territory, than on the street, so don't fling yourself out immediately. Don't go close enough to the windows to be obvious from the outside, but assess the situation. Look for opportunities to secure your location, and check exits. Call the emergency services or building security for help if this is possible and appropriate. Only if all else fails, get out.

⚡ POWER TIP
Don't go for the roof

Resist the urge to follow almost everyone in films who's trapped and seems inevitably to go upwards. In a high building this will reduce your chances of getting out safely (unless you have a helicopter parked on the roof).

Getting out

When it becomes apparent that you will be better off away from the location, the first essential is to get out safely. If you are several storeys up, get down to ground level if at all possible. Check the availability of fire escapes and other means of exit. When you reach the ground floor, find the least obvious means of escape. Look for windows, hatches, maintenance exits and similar opportunities to get out of the building without strolling into the arms of those who are attacking.

One possibility in a commercial building is the service ducts, but don't put too much credence on the way people in movies seem to be able to crawl anywhere in a building in the ductwork – it often isn't possible. Watch out particularly for grilles, weak flooring that won't support your weight, glass and electrical equipment that could provide a hazard inside the ductwork. Never crawl blind through a duct. If the building is breached and you can't get out, a duct can also be an effective place to hide until the coast is clear.

As far as you can, check where the main concentrations of attackers are and look for opportunities – perhaps a back alley, or a window that leads on to a concealing piece of shrubbery – to slip out unnoticed. If there is any gunfire outside, avoid the side(s) of the building where this is happening.

Once you are outside, if the attackers aren't in uniform and are dressed in fairly similar clothing to yours, it may be possible to blend in, rather than make it obvious you have come out of the building. Be prepared to change your clothes as much as you can to look less conspicuous. If you do look much like everyone else after this process, gradually move to the periphery before slipping away. If you stand out from the crowd, try to make it look as if you are coming from a different direction to the building that is under siege.

Once out, unless already pursued, don't run. This is a surefire way to draw attention to yourself. Walk confidently, as if you belong. Only run if blending in with the crowd fails.

Another 'if all else fails' situation is attempting escape out of a high upper-storey window. Although people have survived falls from much further up, a drop of anything over 4 metres is likely to put your life in danger – and even if you survive, you may well break a limb or do other damage and leave yourself at the mercy of those attacking the building – you must try to minimize the distance you fall.

The main essential for surviving an exit from a high place

is to reduce the amount of force with which you hit the floor. If there is anything soft you can use to cushion, your landing site – pillows, mattresses, cushions, etc. – throw them out of the window. Then cut down on the force with which you will impact by reducing the distance you have to fall. Never jump out of a window – hang from the windowsill before letting go: you have already reduced the distance you will fall by your own height.

Look for material you can use as a rope to get down as much of the distance as you can before falling. Offices will be plentifully supplied with cabling (put on gloves, or use some soft material over your hands to prevent friction burns). Look out for ropes, blankets that can be knotted into a rope, or anything else that will support your weight. Fix your means of support firmly to something that isn't going to go out of the window after you. Even if you are using a rope, lower yourself from the window ledge first – the rope may break, and it will help to lower yourself gently on to it, rather than drop with a sudden jerk.

Even if time pressure is tight, make a clear survey of your landing area. Leaving aside any soft items you can throw out to land on, a car roof or dustbin bags could help break your fall (though with the rubbish bags there is always a risk of sharp objects inside). Watch out for railings and tree branches that could impale you. Smaller shrubs should help break the fall a little, though it's worth making sure you have something robust between you and the bush. Wear any padded clothing you've got. If you've a motorbike helmet or a cycling helmet, put that on – otherwise, make a protective turban out of soft material.

If you run out of makeshift rope, the rope breaks, or you have to drop directly from the window ledge, push yourself away from the side of the building with your feet. If there's enough of a fall, try to spin round to face away from the building as you do so. Bend your knees and hold your hands up so your arms protect your head. At the moment of impact, let your knees bend even further then roll on to your side and finally to your back.

Leap of faith

Where the only option is to leap to another building – usually from the roof of one building to another – bear in mind that the movies exaggerate what's possible. Even with a good run-up, you are unlikely to be able to jump more than about 3 metres. A typical city street could easily be 12 to 15 metres across. Only a leap between very close buildings can be risked. The building you are aiming for must have a clear roof (you don't want to be impaled on an aerial or to crash through a skylight) and should ideally be slightly lower than your building, though no more than a couple of metres lower. You will need 15 to 20 metres for the run-up. Try to land on your feet, bending your knees and rolling over on to your side with your head protected by your arms.

However, the chances of getting all these conditions just right are very small. Unless you have no other option, it is rarely a good thing to head upwards towards the roof. This is very much a last resort.

Basic self-defence

This isn't the place to teach self-defence – consider taking appropriate classes. If you haven't done so, the first line of defence is to get away. Don't fight back unless there is no way to get out. If all else fails, act quickly – don't give the attacker time to think. Hit soft targets wherever possible. Don't punch the jaw – you are likely to end up with a broken fist. Use the flat of your hand, your elbow or your knee if possible, rather than a fist. If you are grabbed and your head is near your attacker's, try to smash your head into their nose. See overleaf if the attacker has weapons.

Firearm attack

There is much debate about the merits of carrying firearms for your own protection. In many countries (the UK included) it is illegal to do so, or in fact to carry any form of weapon. There is also good evidence that attempting to defend yourself with a firearm is only likely to escalate a dangerous situation. Don't do it. If threatened with a firearm, forget what you have seen in the movies – it is virtually impossible to disarm someone with a gun before they can shoot you, certainly without a huge amount of expertise on your part. If given a chance to run, get as far away as possible as quickly as you can. Put solid objects between you and the gun – a building, a car, a fence, trees – and plenty of distance. Note that a bullet can travel easily through some of these obstacles (including the body panels of an ordinary car), but at least they make you harder to spot.

Knife attack

As with guns, it is illegal in many countries to carry knives, and to attempt to use one to defend yourself, unless you are expert in handling it, could result in your assailant taking it off you and using it against you.

If threatened by someone with a knife, the best attitude is simply to give them what they ask for. It is all too easy to be seriously injured or killed in a knife attack. If it looks as if you are going to be attacked anyway, try to circle out of range of a slash from the blade – get as far from the knife hand as possible. Use a briefcase, umbrella or other large carried item to hold off the blade as you try to get clear.

Mass panic

In many circumstances it can be helpful to be in a crowd – you are less likely to be attacked, for example (though watch out for pickpockets). But should panic break out, something that is more likely under the pressure of extreme weather, the crowd itself can become a deadly environment. Look for ways to get out – try to work your way to the edge of the crowd. The biggest danger is being pushed over and trampled. Do whatever it takes to keep upright. If you find yourself being pushed towards a wall, or worse a sheet of glass, slip sideways to avoid the obstacle. Try to keep some space around you – folding your arms in front of you might help.

If you do fall, curl yourself up, bringing your head down towards your knees and clasping your hands over your head. Crawl or slide yourself towards any obstacle – something that will force a break in the crowd and that will give you a lever to pull yourself up with. If you spot a gap, you may be able to use another person to help you up, but be wary of pulling them over, or of kicks from someone trying to protect themselves.

⚡ POWER TIP

Riot camouflage

A crowd can be dangerous without turning into a riot, but should a riot break out, you are at risk if you appear different from the crowd. If at all possible, as soon as you see or hear signs of trouble, get away. Don't get involved. If you've no choice, try to blend with the rioters. Make yourself look as unobtrusive as you can. Move with the rioters until a chance comes to slip off at an angle, then get away.

Surviving insects

As temperatures rise, many temperate areas may see a resurgence of insect-borne disease, such as malaria. If the news media alert you to an increased risk of infection, take the same precautions you would expect to take when travelling to a country where such diseases are common.

It's best to wear long-sleeved clothing and long trousers,

particularly after sunset when attacks are much more prevalent. Also cover any exposed skin in insect repellent. Count thin clothing that an insect could bite through as exposed skin. Use insecticides indoors and set up screens on windows and doors. If the room you are sleeping in isn't properly screened and swept for mosquitoes, sleep under a mosquito net impregnated with insecticide.

Vaccinations are available for some insect-borne diseases if they should break out in your country, including yellow fever and Japanese encephalitis. Although there are anti-malarial drugs, they need to be taken for a period before exposure, during exposure and a period after, so while useful for a short-term trip, they don't give the benefit vaccination does of preventing attack over a long period. In principle some anti-malarial drugs can be taken for five years or more, but get medical advice before taking anti-malarials over a long term.

Basic first aid

With the impact of climate change, it's possible that you will have to be more self-dependent because emergency services take longer to get to you – or don't come out at all. Whether or not this is the case, a basic knowledge of first aid is valuable for everyone. It's best to get this from a course – many countries have first-aid courses available for free or for a small charge – but below are some essentials if you haven't had the chance to go on a course yet. Note that these are normally literally first aid, to tide things over until you can get professional help – they are not a substitute for proper medical care.

If there are lots of people around, make sure that the best qualified to help are assisting the person in trouble and get everyone else back from the scene. An audience is no help, upsetting for the victim, and will get in the way. If practical, though, it can be reassuring to the victim if one person can keep in communication with them – reassuring, calming – while another undertakes any medical help. Make doubly sure that someone has rung for help – it's easy to assume in a crowd that someone else has done this.

If you are the best person to provide help, first assess the situation. Don't panic. Don't move the person who is hurt unless there are life-threatening reasons why they cannot stay where they are: moving someone with certain types of injury can make things much worse.

Talk to the person first before touching them. Ask 'Are you OK?' in a firm voice. Call the victim by name (if you know it) several times, looking for a response. Then the essentials to check for are breathing and bleeding.

Breathing checks

Check that the victim is breathing normally. It can be difficult to see if someone is breathing – look for movement of the chest, listen for breath from mouth or nose, see if a mirror or cold piece of metal held very close to the mouth and nose mists up. Provided there is no possibility that neck or spine have been injured, if the casualty is lying on their back, you can gently tilt back the head and position the jaw forward to clear the airway.

POWER TIP
Neck and spine injury

If there is damage to the neck or spine, any movement could make things worse or even result in death. Suspect this if there has been a blow to the relevant parts of the body. Lack of feeling below the neck or pins and needles in the extremities could indicate back or neck injury.

If there is any blockage in the mouth, try to clear it without pushing anything down the throat. (See page 192 for choking rescue, if this fails.) If you are confident there is no neck or spine injury, and the casualty is breathing, move them into the recovery position.

The recovery position

If you are certain there are no neck or spinal injuries, or cuts and breaks in places that would make it impractical, and the casualty is breathing, get them into the recovery position, which will help prevent choking on any vomit, the tongue or fluids. Roll the victim gently on to their side, bringing out the uppermost leg and arm in L shapes to support the weight. Turn the head to face the same direction as the sticking-out arm and leg, and gently angle the head back with the chin carefully pulled forward to keep the airway open. Keep an eye on vital signs. (If there are no breathing difficulties and the victim appears to be suffering from shock use the shock recovery position instead – see page 199.)

No breathing?

Once someone stops breathing there is very little time left. Brain damage can occur very quickly – in four to five minutes. If there is no blockage to the airway (see 'Choking' rescue, page 192), make sure that the head is tilted back and the jaw dropped to open the airway. Take out any loose dentures and loosen any tight clothes around the top of the body. Then undertake cardiopulmonary resuscitation (CPR) (see below). Older manuals suggest checking for pulse before giving CPR, but the most recent guidance takes the view that it is so difficult for a non-professional to find a weak pulse that this stage should be omitted. It was thought that chest compressions could stop a feeble heartbeat, but there is limited evidence for this.

> ⚠️ Note that done incorrectly, chest compressions can cause fatal damage. It is essential that you are trained to give CPR – if you aren't trained, only attempt resuscitation if you are instructed to do so by emergency services, or if there is no way to get assistance from someone who is trained.

Squeeze the victim's nostrils closed and place your mouth over their lips. (If you have a mouth protector in your first-aid kit, use it.) Blow twice in fairly quick succession, watching out for the chest rising. If the chest doesn't rise when you blow, check for obstructions. (For a baby or small child, put your mouth over the mouth and nose. Don't breathe too hard, or tilt the head back too far.) Alternate mouth-to-mouth with chest compressions (see below). Keep going until the casualty is breathing normally – then place in the recovery position. If for any reason you can't use mouth-to-mouth, you can cover the mouth with your thumb and use mouth-to-nose instead.

For chest compressions, make sure the casualty is lying flat on their back. Place the heel of one hand in the middle of the chest, between the nipples, fingers pointing towards a nipple. Put the heel of your other hand on top of the first so the hands are overlapped and parallel, keeping your fingers away from the chest.

Keeping your arms straight, push smoothly and firmly down, then relax, at a rate of about 100 presses per minute for 30 compressions, pushing down 4 to 5 centimetres and allowing the chest to return to position each time. Then go back to mouth-to-mouth for two breaths, continuing to alternate mouth-to-mouth and chest compressions until help arrives. With small children use one hand only for the compressions, pushing down around one-third of the depth of the chest, and with infants (under twelve months) use two fingers only.

Electrocution rescue

As with all emergencies, if phones are operating, call for medical help as soon as possible.

❒ Check for water on the floor – don't approach if the floor is not dry.

❒ **Don't touch the victim** until you have established that the power is off.

❏ Switch off the power. Isolate the area if possible at the mains/consumer unit. Otherwise switch off/unplug at the socket.

❏ If it is not possible to cut the power, try to knock the victim free using an insulating item like a piece of wood, or pull them indirectly using a dry towel or other piece of material.

❏ Don't move the victim more than is necessary – there may be other injuries than those directly from the electrocution.

❏ Check for breathing. If necessary, give CPR (see page 190).

❏ Keep the victim warm and quiet.

 Note that this advice is for ordinary mains voltage. High-voltage cables, such as those on electricity pylons or used to power trains, should not be approached under any circumstances. Do not even try to help a victim using a stick etc. The voltage is sufficiently high that it can jump a gap, and can cause a deadly shock even through a relatively poor conductor like a piece of wood. It is essential to get such power lines deactivated before attempting any rescue.

Choking rescue

As with all emergencies, if phones are operating call for medical help as soon as possible.

⚡ POWER TIP

Hold back on Heimlich

If someone is choking, it is important to act quickly. Make sure they have no loose dental work in their mouth, and check for anything blocking the mouth and throat. Do not copy the films and go straight for the Heimlich manoeuvre (where you stand behind them and pull sharply into their stomach) – this can cause damage and is a last resort. Be careful to distinguish choking from a heart attack (a heart attack victim will have chest pains and possibly arm pains, and the victim's struggle will be focused on the chest rather than the throat).

First check for and remove anything blocking the victim's mouth or throat, and any loose dental work. If that has not cleared the airway, administer three or four sharp blows on the back between the shoulder blades with the heel of your hand (check for dislodged matter after each slap). Adults should ideally be sitting down and bent over so that the head is lower than the knees. If the adult is unconscious, lay them on their side, and support their chest. Children or babies should be laid across the lap with the head hanging down. Tailor the slap to the child – the smaller they are, the gentler you need to be.

Only if this fails use a compression technique, as this can be highly dangerous. If the casualty is unconscious, use a similar approach to the CPR compression (see page 190), but with the heel of your hand between the navel and the breast-bone. Try up to four thrusts, which should be quick and upwards, towards the centre of the chest cavity.

For conscious victims, use the Heimlich manoeuvre (but see above). Stand behind the victim and place your fist, thumb inwards, between the navel and the breastbone. Put the other hand over the first and sharply pull inwards and upwards. Repeat, up to four times in all.

If the victim is unconscious, after clearing the airways lay them in the recovery position (see page 190).

Burns rescue

As with all emergencies, if phones are operating call for medical help as soon as possible.

If the victim is actually on fire, get them to roll on the ground. Prevent them from running – this will just fan the flames. Cover the flames with a thick piece of clothing or rug to deprive the fire of oxygen and put the flames out. Get any still smouldering clothing off as soon as possible.

Get the victim away from whatever has caused the burn. Apply cold clean water copiously, preferably for at least ten minutes. Remove any jewellery or clothing from the affected area, but do not try to get clothing that is stuck to the burns away. Cover the burned area with a clean, lint-free material.

It's very important not to burst blisters – these are a natural protection – and don't use cream, ointment, lotions or other once-recommended treatments like butter. The covering should not be sticky: don't use plasters. Give burn victims sips of water to counter fluid loss, unless they are in shock (see page 198).

Bleeding rescue

As with all emergencies, if phones are operating call for medical help as soon as possible.

If the wound is on an arm or leg, lift the limb so it is higher than the heart, so reducing the pressure that is pushing out the blood. A small cut should be treated in the usual way – cleaned and covered with a plaster. It's OK to use something like cotton wool in the cleaning process, but don't use it in the dressing – this should be lint free. In emergency, use a clean cloth or, if nothing else is available, absorbent paper.

With a large cut, cover with a dressing and press on it to reduce bleeding. If nothing else is available use a suitable clean object like a credit card. If the wound is opening outwards, try to hold it closed. If the object causing a major wound – a knife or a large shard of glass, for instance – is still in the wound, don't remove it, as you could make matters worse. Keep up the pressure for at least five minutes to ensure clotting. For safety be prepared to wait until help arrives. Don't try to cut off blood flow by twisting something around an arm or leg (a tourniquet) – this can do more harm than good.

It's particularly important to seal a chest wound,

especially if there is frothing blood around the wound and the sound of air being sucked into the chest. Cover up the wound with a hand or credit card or sheet of plastic and apply pressure to make a seal until you get medical assistance. Keep the victim propped up, half sitting, with the functioning lung higher than the other.

If a large amount of blood is escaping and you can't prevent it, you can use a pressure point to reduce the flow. For an arm injury, feel between the muscles on the inside of the upper arm. Push against the bone inwards and upwards until the bleeding reduces significantly. Hold for up to fifteen minutes, but no more as damage could result. For a leg injury, lift the knee and press on the top of the leg, in the middle of the fold where the thigh meets the groin. Again press until bleeding reduces significantly.

POWER TIP
Proper antiseptic use

Antiseptic should be applied to shallow cuts and grazes to reduce the chance of infection (if you have no antiseptic, soap helps). With deep cuts, use antiseptic around the opening, but don't pour it into the cut itself – use cooled boiled water to wash out the wound.

Try not to come into direct contact with blood – use gloves or plastic bags over your hands to minimize the chance of infection.

Although a nosebleed is less serious than most of the above, it's worth listing the recommended approach, as guidelines have changed so much over the years. Get the patient to sit and pinch (they can do this themselves) the top of the nose on the soft part at the end of the hard cartilage for five minutes. Ask them not to sniff, and to try to breathe through the mouth, not the nose.

Dislocation rescue

A dislocation is a bone that has popped out of its normal socket or location. It's best to immobilize a dislocation and wait for professional help if this is possible. Only attempt to reposition the dislocation if there is no hope of assistance. The aim is to pop the joint back in place, which should be done as soon as possible if you have to take action yourself, as with time it will become harder to do because of swelling and muscle spasms.

Gently feel the area to try to ascertain that this really is a dislocation, not a fracture – it's difficult to be sure (another reason to wait for expert help if available in time). If the joint is a finger or thumb, take hold of the end of the digit and pull outwards – the joint should pop back into position. Once you have done this, tape the finger or thumb to the next finger along to keep it as still as possible during recovery.

The next most common dislocation is the shoulder. Before taking any action, try to determine if this really is a dislocation, as there is another shoulder injury (separation) that can easily be misinterpreted. If the victim fell on to their shoulder, there could be some tearing of the tissue between the shoulder and collarbone – this is shoulder separation. Gently feel around the collarbone. If the flesh feels soft and gives more than usual, this is likely to be separation. Dislocation is less likely from an impact on the shoulder, and more likely to be from a wrench or twist to the arm and shoulder. With separation, strap up the arm and get assistance. With a dislocation, if no assistance is available get the patient lying flat on their stomach on a strong table, with the damaged arm hanging over the edge. Pull down firmly on the victim's wrist until the shoulder pops back. If no suitable table is available, take your shoe off and place your foot in the armpit, then pull on the arm until the shoulder pops back. Use a sling to provide support afterwards.

Hip and elbow dislocations are significantly harder to deal with and you may have to make do with immobilization.

Fracture rescue

Broken limbs are not uncommon injuries, and will benefit from some first aid before professional help is available. Assuming that professional help will be accessible in a reasonable timescale, the main requirement is to stop the broken bones from moving, to prevent both further damage and excruciating pain for the victim.

If, and only if, there is no hope of medical assistance, try to reset the break by applying traction – pulling slowly, carefully and powerfully to separate the broken halves, then positioning the bones back in line. When in position, keeping up the pull until fixed, use a splint and immobilize the break as much as possible. This will be painful – use painkillers if available. Otherwise just immobilize, using a combination of a sling where appropriate and a splint or splints (straight pieces of rigid material like wood) with padding to avoid the splint digging into the wound and a binding bandage to hold it in place. Consult a detailed medical guide for individual immobilization recommendations. Make sure blood is circulating to fingers or toes – danger signs are tingling and blue or over-white colouring. Loosen the bindings if necessary.

POWER TIP
Fractured skulls, backs and necks

Fractured skulls, backs and necks are all beyond first aid. Try to minimize movement. For skull injury, place the victim in the recovery position (see page 190) if movement is possible, allowing any fluid leaking from nose or ear to escape. With potential spine or neck injuries, keep movement to a minimum. If you can use something to minimize movement, for instance a DIY cervical collar to support the neck, do so, but only if you can get it in place without causing any movement in the process.

Broken ribs can't be splinted – keep the victim still and give painkillers. Ensure that the victim has a medical check as soon as possible as there could be internal damage. Broken toes, with the possible exception of the big toes, don't usually require direct treatment, but are very painful – use appropriate painkillers. Still get the victim checked by a medical professional to make sure this is the only damage.

Bite rescue

Bites can cause considerable problems if not properly treated. As always, seek immediate medical assistance if at all possible.

Animal bites (including human bites) are liable to cause infection. Clean the wound carefully and give antiseptic treatment before dressing as for a normal wound. As soon as possible get treatment for possible rabies infection.

Despite everything you've seen on the TV, don't try to suck poison out of a snake bite or spider bite. Get emergency help as soon as possible – if you can kill the snake or spider as evidence of the species do so, otherwise photograph it or at the very least note its obvious characteristics (size, markings, type of movement). Get the victim to lie down with the bite lower than the heart, keep them as motionless as possible, and try to calm them. Wash the bite and apply antiseptic. Loosen clothing, particularly if there is any sign of swelling.

Insect stings, while not technically bites, can still be unpleasant. Use an antihistamine cream or tablets if available (hay fever tablets). If there is any sign of shock reaction, particularly difficulty in breathing, get assistance as soon as possible.

Shock assistance

This is not for an electrical shock (see page 191) but for the impact on the body from an accident. Shock is the body's response to trauma. It limits blood flow to non-essential organs in an attempt to maximize survival chances, but iron-ically the shock itself can kill, particularly if the person suffering shock is weak, elderly or has a heart condition.

Typical symptoms of shock are feeling weak or dizzy, and being unresponsive, having pallid or blue skin that feels cold and possibly clammy, a weak but rapid pulse (typically over 100 beats per minute), rapid shallow breathing and nausea, sometimes with vomiting.

POWER TIP
Finding a pulse

It was once recommended that you only give CPR if you can't find a pulse, but the recommendation now is that amateurs ignore this stage as it can be difficult to do and wastes time. However, it can be useful to find a pulse in a breathing patient to check (for example) for shock symptoms. Rather than feel or listen at the chest or wrist, try to find the pulse in the carotid artery in the neck. Place your index and middle finger (together) gently on the victim's Adam's apple at the top of the windpipe just under the chin, then slide your fingers at the same level towards the side of the neck, off the windpipe on to the muscles. When off the windpipe press slightly more firmly and you should be able to find the pulse.

If the victim is in shock and has no breathing difficulties, don't use the recovery position. Gently manoeuvre the victim on to their back, supporting the head, but not raising it. The legs should be horizontal from knees to feet, bending at the hip, with the feet raised about 30 centimetres off the ground. As with the normal recovery position, don't move the victim if there is any risk of their having suffered back or neck injury.

Loosen any tight clothing and keep the victim's body heat in using blankets or a sleeping bag, but don't add any extra heat from heaters or other sources (this will bring blood to the skin, taking it away from vital organs). Don't give any fluids by mouth. Try to get medical assistance as soon as possible.

Childbirth assistance

Giving birth isn't an illness or wound, but it is a process that few of us have experience of handling unassisted.

Reassure the mother-to-be and make her as comfortable as possible. Loosen any restrictive clothing. The early stage of contractions can take a day with a first child – be prepared for a long haul. Keep the mother warm. Try to make her comfortable with her knees in a drawn-up position. Discourage too many spectators – the mother isn't a freak show. Keep things as hygienic as possible.

Once the waters break, sending the water-like amniotic fluid pouring out, the next phase has begun. During contractions, get the mother to hold her knees and pull her head towards them, bearing down. Between contractions she should relax as much as possible.

When the baby's head appears, support it, using a clean towel to keep it away from the anus. After the widest part of the baby's head is out, ask the mother to stop pushing and switch to panting. Clear any mess from the baby's face and make sure it isn't caught around the neck by the umbilical cord. Gently lower the baby's head so the higher of its shoulders (it should have rotated if necessary) is able to slip out – then return support to help the other shoulder out. It should now be possible to lift the baby out and put it on the mother's stomach.

Clean the baby's face again. It will probably cry and should be breathing by now. Don't slap it to get it to start breathing – if necessary use very gentle mouth-to-mouth-and-nose (see page 191). Wrap the baby up to keep it warm and lay it face down so that any remaining fluid can drain.

Don't pull on the cord – wait for the afterbirth to be naturally expelled (this can take half an hour or so). Give the mother a towel to help staunch any residual bleeding. Don't try to cut the cord – leave this to the professionals.

Cute isn't safe

Be wary of wild animals, particularly if there is a water shortage or food shortage for the animals. It's not that a rabbit is going to try to eat you (though a fox might), but it may bite or scratch in self-defence.

⚠️ *In many countries (unlike the UK and Australia), rabies is common in many wild animals. Keep well away from racoons in North America, for example. If you are bitten by any animal (including unknown dogs), or even smeared with their saliva, it is worth getting medical help in a region with rabies on the loose.*

Resources

Fearsome Fido

A reasonably large dog can be a very helpful aid to home security. Many intruders will be concerned about taking on a big dog, and even the friendliest of dogs can turn nasty if they feel their pack (the family) is under threat. Like all pets, a dog adds to the burden of the household when you are facing the difficulties of climate change – a large dog will consume a fair amount of food and water – and a dog requires a considerable amount of care. Contemplate taking one on only if you are prepared for the requirements of looking after it. But given those negatives, a suitable dog will add to your security. (Although some small dogs like terriers can be fierce in temperament, the dog does need to be of at least medium size to have the appropriate psychological impact on an intruder.)

The armoury

The decision whether or not to have weapons in your house is a complex one. Many countries, including the UK, allow only certain types of gun to be kept, and then only by a licensed owner, with the weapon held securely by in a locked gun cupboard. Low-powered airguns are usually less regulated – in the UK anyone over thirteen can use an airgun on private property, but they cannot be used in public places, or carried unless unloaded in a secure case.

Apart from complying with the laws of your country, you ought to take into account that a weapon like a gun, if it falls into an attacker's hands, will do more harm than good. In fact the people most often killed or injured by a gun kept around the house for protection are the householders. You might consider keeping an unloaded gun, so that it can be used to scare an attacker without risking harm to yourself, though bear in mind that if it is stolen it could easily be loaded and used against someone else. Guns, and other projectile weapons like crossbows and longbows, are most likely to be of benefit when the attackers are outside the building and trying to get in – then they can be used to your advantage without the risk of the attackers getting their hands on them. If possible, try a warning shot first – many attackers will be scared off.

Most houses already have a good supply of knives in the kitchen, but again going into a fight using a knife presents a strong danger that a more experienced attacker will get the knife off you and use it against you. It might be worth keeping your kitchen knives tucked away in a drawer, rather than out on view in a knife-block, advertising their presence to an intruder (although there are no statistics to prove the effectiveness of this precaution).

One way to deal with attackers who are outside the building trying to get in is to resort to the old siege favourites of boiling water or hot oil, poured from an upper storey. You need to be careful that the attackers don't have guns or other projectile weapons, as you will be quite exposed when doing this. Take extreme care getting the hot materials to the upper floor, as there is significant risk of hurting yourself or others in the house.

The traditional improvised club in the form of a baseball bat or cricket bat isn't a bad self-defence weapon if the attacker is already in the house. Don't telegraph your swing by taking the club too far back before hitting out – you could give your assailant too much time to respond. As soon as the club has come into contact, pull it sharply away from the attacker. If you haven't done much damage and you leave the end within easy reach, you could turn your defence into a wrestling match for ownership of the club.

A common household alternative to the club is the gardening implement, many of which can be very unpleasant when wielded against an attacker (in the history of warfare, many fighters have been armed only with agricultural implements). The one problem here can be access, as garden tools are often kept outside the house. Make sure you aren't offering these on a plate to attackers – if they are kept outdoors, lock up the shed or garage where they are stored.

Evacuation practice

When it is time to get out of the house quickly – whether to escape fire, attack from outside, floods or for other reasons – it is too late to try to tell your family what to do as you go. Take everyone through the basics of getting out before any emergency occurs. Make sure everyone knows how to open any double-glazed windows, and how to cope if a window is locked, or can't be opened enough to get through: use a hammer (preferably a specialist device – see the appendix, page 261) to smash the glass, hitting it at the corner of the frame first. You may also break through by throwing a suitable heavy object, but be aware that it may bounce back, so don't stay in its path.

Consider what small children whose bedroom windows are kept locked for safety reasons should do. If you keep the keys to external doors somewhere out of sight – a sensible idea for security reasons – make sure everyone old enough to be responsible knows where the keys are.

In case you have to get away from the house quickly, or some of your family aren't at home when you evacuate, arrange a meeting place. It should be somewhere that is sheltered in case family members have to wait several hours and should be either very public or well hidden.

First-aid training

One of the best resources you can have around the house is someone trained in first aid. Encourage everyone in your family from teenagers upwards to take a basic first-aid course. They are often available free or at very small charge through your government, health service or charities like St Johns Ambulance. Knowing the basics, from CPR to treating a bleeding wound, is valuable at any time, but could become a life-or-death essential if global warming's impact reduces the immediate availability of medical services.

CHECKLISTS

Basic first-aid kit

Make sure you keep a simple first-aid kit in the house and in the car. It can be tempting to make use of items from the kit for everyday essentials – for example, borrowing the scissors or using a sticking plaster for an everyday cut. If this happens, restock it at the first opportunity. When it comes to a real emergency, you don't want to find half your kit is missing.

Painkillers (paracetamol and ibuprofen) .... ❑
Never exceed the dose on the packet, but if the patient is in serious pain, it is acceptable to give the full dose of both paracetamol and ibuprofen, increasing the pain relief. Don't mix any other types of painkiller. As with all medication, don't take painkillers if they cause a negative reaction, and do not use after the expiry date.

Mixed set of sticking plasters (minimum 20) .... ❑

Sterile dressings (minimum 4) .... ❑

Roll of adhesive fabric tape to secure dressings .... ❑

Crepe and tubular bandages for support .... ❑

Triangular bandage to make into slings .... ❑
You will need some safety pins to use with this.

Cotton wool (for cleaning wounds only) .... ❑

Antiseptic (cream and wipes) .... ❑

Antihistamine cream (for stings and bites) .... ❑

Scissors .... ❑

Tweezers (for splinters) .... ❑

Small mirror .... ❑
Can be helpful for self-treatment, and for checking for breath.

Eye-care items (eye bath and dressing) .... ❑

Emergency evacuation kit

It may be that for whatever reason you need to get out of your home and survive for a few days with no prior warning. Perhaps your house is under attack and the best thing is to get out and take cover elsewhere. Having a kit that you can grab and take with you at a moment's notice seems like overdoing the planning, but you may not have time when all hell breaks loose. The difficulty here is to achieve a balance between what you really could do with and what you can carry.

Food and water for three days .... ❑
Keep this as portable as possible. Limit the food to essential rations and water to 2 litres per person per day.

Water purification kit .... ❑

First-aid kit (see page 204) .... ❑

Torch (wind-up) .... ❑

Waterproof (lifeboat) matches or cigarette lighter ..... ❑

Utility knife (Swiss Army knife, for instance) .... ❑

Warm clothing and wool blankets/sleeping bag .... ❑

Battery (or better, wind-up) radio .... ❑

A few dustbin liners .... ❑
Useful for insulation and waterproofing as well as containers.

Waterproofs .... ❑

Small tent or plastic sheeting .... ❑
Modern tents pack up very small and can be reasonably easily shared as a load between a few people.

Mobile phone .... ❑
Doubles as an emergency light, and an essential for communication if the networks are still operating.

A secure home

It is almost impossible to make a building entirely secure, but you can reduce the vulnerability of your house to attack by building up its safety features

Good-quality external locks .... ❑
Make sure all external doors have five-lever (or more) mortise deadlocks.

Protected patio doors .... ❑
If you have patio doors make sure they have locks at both ends so they can't be levered off the track.

Window locks in place .... ❑
All windows that there is any chance of someone getting through should have locks, and be kept locked when not in active use.

Survey for other points of entry .... ❑

Check the outside of your house for other ways in apart from doors and windows. Make sure that they are protected.

Consider metal shutters over windows .... ❑

If you don't have metal shutters, look into fitting them.

Make it possible to isolate unprotectable rooms .... ❑

Fit high-quality doors and locks at the entrance to the main house from rooms like conservatories that can't be sensibly protected against break-ins.

Disaster plan

Make sure that you are ready before the problems hit.

Everyone understands how to get out safely .... ❑

Keys are kept securely in a location everyone knows .... ❑

At least one first-aider in the house .... ❑

Meeting place arranged in case of evacuation .... ❑

Kitchen knives safely away in a drawer .... ❑

Any weapons in the house are secure but ready to hand .... ❑

SIX

KEEPING OUR HUMANITY

It's easy to become obsessed with the physical aspects of survival in the difficult circumstances that are likely to prevail as a result of global warming. But we have to deal with inner pressures too.

What to do ...

... to cope with stress

... to relax

... to be entertained without technology

... to solve problems creatively

... to enjoy pure creativity

Coping with change

The future that faces us may be unpleasant, thanks to global warming. Frequent power cuts. Difficulty obtaining food and water. Storms and floods. A breakdown of society, where just walking down the street could be a high-risk activity. Most survival guides limit themselves to what's required to stay alive, but that's a miserable, short-term view. There comes a point when surviving isn't enough – there needs to be something more. This section looks at how to keep your humanity and to make more of your global-warming survival than just scratching a living.

Humans are physically fragile animals. We have survived as long as we have by adapting and by using our brains in order to go far beyond the basic biological limitations of our bodies. We need to make the most of our human capabilities and the edge that they give us.

One essential is being able to deal with the stress of change. The biggest stresses in everyday life – moving house, getting married, bereavement – are triggered by change. The impact of global warming is inevitably one of change in our lives, and that means increased stress. Stress isn't inherently good or bad. We all need some stress to give us an edge, to push us to achieve. But too much stress leads to serious health problems. This is particularly true of negative stress that can't be countered. If you feel out of control, stress is particularly likely to cause you problems. The practical guidance in the previous chapters will be a big help in moving yourself away from feeling out of control, but even so it is a very practical addition to your survival kit to have simple techniques to hand to help keep stress at bay.

Stress management sometimes gets lumped in with spirituality as something more touchy-feely than practical – but it doesn't have to be like this. The stress-management guidance in this chapter is just as solid and practical as the chapters on dealing with blackouts or extreme weather. The danger here is less visible, but it is no less significant.

There are two levels at which you can deal with stress: the physical and the emotional. On the physical side, basics like eating and sleeping well, and taking more exercise, are big contributors. These may be obvious, but that shouldn't undermine their significance.

There are also physical treatments for stress which encourage the brain to trigger the release of stress-relieving chemicals into the body – massage, for example.

It's not enough, though, to consider the physical. A large degree of our response to a stressor like the impact of global warming is dependent on our emotional state and self-image. Being depressed and unhappy magnifies the impact of the stress. Being in a good state emotionally has no impact on global warming itself, but can make the difference between it being an impossible burden and an inspiring challenge. Self-confidence is also hugely important. Strong self-esteem is a powerful weapon when fighting off the effects of stress. Research has shown that people who have apparently stressful jobs – company directors, successful self-employed people, surgeons, air traffic controllers – are much less likely to succumb to stress-related illnesses than those in apparently low-stress jobs. This is both a matter of self-esteem and of being in control. The more we feel that we are in control, the less stressed we are likely to be, even in a difficult situation.

The more we feel that we are in control, the less stressed we are likely to be.

Stress is a negative aspect of the impact of climate change, but as human beings we need more than simply to overcome the negative and exist. It's important when looking at a new life brought on by global warming that we find the aspects of living that go beyond survival. One essential contributor to our humanity is creativity.

Whether it's the creativity of finding a solution to a lack of water or the creativity required to tell a story, you will find making the best of your natural creative capabilities is an essential part of surviving climate change, so a significant part of this chapter is about supporting your ability to think creatively.

However valuable creativity may be, though, there is little point in covering the subject if there's nothing practical that can be done to change our capabilities. Luckily, over the past fifty years, a lot of work has been put into developing ways to improve on natural creativity. For our purposes, two aspects are particularly important and will feature in the practical advice in the 'Solutions' section: environment and techniques.

The environment in which we try to solve problems and come up with ideas is relevant for good physiological reasons. If you are under real pressure or distracted, your brain isn't good at forming the new pathways it needs to generate an idea. Instead, it makes use of well-trodden paths, churning out the same old solutions. But these may no longer be applicable in a changed world – so getting away from pressure and distractions, taking a moment to think, can be crucial.

The second aspect, creativity techniques, involves a series of short exercises designed by psychologists and other experts as ways of pushing us into thinking differently. Such techniques can be extremely valuable to kickstart a new way of looking at things and to come up with solutions to the problems that global warming throws up.

Creativity is beneficial in overcoming obstacles – but it's more than that, it's a positive act. If you can be creative, come up with a new idea, solve a problem, then it gives you a good feeling. And that can't be bad when you are facing the realities of climate change.

Solutions

Building self-esteem

Self-esteem level is an important contributory factor to your ability to cope with stress. If your self-esteem is low, you are much more likely to succumb to stress-related illness. One of the undermining factors that keep self-esteem low is the diminishing spiral that begins by thinking 'I never achieve anything.' The result is that you feel bad about not achieving. Because of this, you get stressed and achieve even less. It's as clear a case of positive feedback as any of the systems influencing global warming.

A very quick exercise can have a surprisingly powerful effect on self-esteem. Spend a couple of minutes jotting down a handful of small achievements you have made today. However bad a day you've had, you should be able to find something positive to say – force yourself to generate at least three; don't take 'no' for an answer. Repeat this exercise each day for a week. Stick to small achievements for this first stage. No one is going to have a big achievement every day, but everyone has a series of small successes that prove the fictional nature of the destructive view that everything about your life is terrible and you never succeed at anything.

It might seem that small successes – they might just be 'I got out of bed on time' or 'I told my children a bedtime story' – are small beer compared to the problems that global warming is throwing at you. It doesn't matter; much lack of self-esteem derives from an imagined bleak picture that *everything* is going wrong for you. Realistically this can't be true – and proving it to yourself can really help.

As a separate exercise (do it on a different day), sit down with a notepad and note down some of the occasions in your life when you've had a big success. Include events that are important only to you as well as things that seem important to the world at large. It could be success in exams and education, getting a job, getting married, having children, the first time you did something that was hard to achieve (driving a car or beating someone at chess, for example). As far as the world is concerned, these achievements might not be large, but they do need to be significant to you.

When you've got together a list of successes – it doesn't matter whether it's a handful or a great string – spend some time thinking through what it felt like at the time. Re-live the moment when you realized you had succeeded. Don't feel guilty about enjoying your success; you deserve it. Remembering a moment of triumph has a surprisingly strong influence on your current feeling of well-being.

By working on your self-esteem you can add hugely to your ability to stand up to stress and to deflect it; all the evidence is that those with high self-esteem are better able to cope with stress. Don't worry if this exercise sounds artificial; it still works. You could repeat it, perhaps yearly, looking back over your achievements of the last year.

One last quick self-esteem exercise. Spend five minutes thinking about how your week is divided between doing things for others and doing things that *you* really want to do. You may find a frightening lack of time for yourself. Make sure you get some time in your week that is yours to do with as you wish. This is particularly important if you are self-employed.

POWER TIP

Self-esteem isn't self-centred

The tendency in the latter half of the twentieth century was to increasing self-centredness. People left their families to 'find themselves' or ignored responsibilities in pursuit of pleasure. There is something of a backlash now, because the cost of this cult of the individual has been misery from a breakdown of social values and a realization that the pursuit of success for its own sake isn't particularly rewarding. In looking for stress relief you need to get some space for yourself, but you also need to look outwards as well as inwards. What's needed here, as so often in stress relief, is a balance. Not placing yourself above everything else, but equally not ignoring your own needs.

Finding some time to be you, and to do what you want, is a great way to relieve stress. Most of us are out of balance in this respect. But remember the need to look outside yourself as well. It's easy to put off finding some time for yourself – particularly when trying to deal with something as intrusive as climate change. Make sure that you succeed.

A stress workout

Exercise reduces physical tension and brings down levels of stress chemicals. It also builds up the body, helping general fitness and ability to cope. The impact of global warming may mean you have to undertake more physical effort than previously – but many of us still don't get enough exercise, and exercise is an essential part of a stress reduction programme. In a sense, doing exercise isn't the problem here, it's sticking with it. Most people who suddenly decide to take exercise don't keep it up. Try this three-point plan.

1. Self-motivation. Find a driving reason to exercise (go for the gut, like 'staying alive for my children'). Make sure it is at the forefront of your mind when you decide how to use your time.

2. Choose something you enjoy. This may seem self-evident, but many people choose a form of exercise because it's trendy (the gym) or because it might be career-boosting (golf). Find something you really enjoy.

3. Add value. Get together with friends and make exercising a social event, or choose an activity where you can wear an MP3 player and listen to music or speech radio or audio books, or use the time to learn a language.

Typical choices are swimming, cycling, running or gym routines – avoid sports that don't involve continuous activity. If, like me, you find many of these boring, don't ignore walking. A fast walk can be very effective exercise (especially if hills are involved) – it involves less risk of joint damage than jogging, and burns more energy per kilometre than cycling. What's more, walking is practical. You can get something directly out of a walk, whether it's taking in great scenery, collecting firewood, exercising the dog or posting a letter. I hate exercise for its own sake – having a practical goal doubles the value.

POWER TIP
Integrate with normality

If you have trouble getting started with exercise, integrate it into your normal life. Get off the train or bus a little sooner and walk the rest of the way to work. Don't take the lift; use the stairs.

Plunging into heavy exercise can cause health problems. Get some guidance if you are in any doubt.

Walk some more

Walking isn't just a good form of exercise, it scores highly on a number of counts. It isn't challenging – anyone can do it – it doesn't make you look odd, and unlike most exercise it isn't mind-numbingly boring (if you approach it the right way).

If possible walk somewhere you can take in the natural stress relief of the countryside (or at least a park) – fresh air, greenery, lack of traffic. But if you can't get to the countryside, at least get outside and really take in your surroundings. If you are walking past a row of shops, don't just look in the windows, look up – see the rest of the buildings. Observe the people around you and what they're doing. Remember to use suitable footwear – trainers may not be your usual style, but they're much better than typical office shoes.

There are two approaches to stress-relief walking. You

can either deliberately keep all your thoughts at bay, or let them work through. In the first approach, focus on your surroundings. Don't let your thoughts wander back to work or home problems. Imagine you are an artist or writer or composer and want to capture your surroundings – take them in, both in depth and in overview. If there are people around, take an interest in them (not too obviously) – everyone is interesting.

The alternative approach is to pick whatever's going through your mind most at the moment and face up to it. Whatever is the big problem at work or home. Just let the problem and any surrounding facts slosh about in your mind. Don't make a heavy effort to find a solution – let things happen at their own pace.

Walking gives you the triple benefit of exercise, fresh air and an opportunity for your mind to work in a very different way. As an added bonus it's a defence against stressors because you're usually out of reach (don't take your mobile). Make it happen. Fifteen minutes is a sensible minimum that you should be able to do several times a week – half an hour would be even better.

Sharing chores

Spend a minute or two thinking through a series of typical days. Look out for regular activities that you don't enjoy but you always end up doing – there may well be more of these as climate change means you have to fend for yourself more. These chores could be at home (getting up with the children in the morning, washing up, ironing, growing food, collecting water, dealing with the rubbish) or at work.

Look at ways that you can share these tasks around more. Sometimes it will be just a matter of swapping a chore – doing someone else's chores can be surprisingly pleasant compared with doing your own. It may be necessary to renegotiate your division of labour, but if this is the case, go into it positively and lightly. Any attempt to charge in demanding rights is liable to wind everyone up the wrong way.

The division of a particular chore doesn't have to be equal. It might be that you enjoy the job despite its mundane nature but don't want to do it every time. In such a case, being given a surprise break every few weeks can be just as beneficial as a rota, and much less bureaucratic. Sometimes, if you are the only one doing a dirty job, it could be because your view of what is important doesn't fit in with everyone else's. If this is the case, try stopping doing it. If you find you can manage without it, fine. If other people miss it, encourage them to join in from now on.

Chores are small activities that don't seem particularly significant. However, if it's constantly assumed that you will do the dirty jobs, you will find it depressing and stressing. Sharing the chores around makes a lot of difference. You could use a rota, and sometimes this is inevitable, but try to operate without one first. Few people enjoy the rigidity of a rota – keep it for situations where the task won't get covered otherwise.

Laugh!

As we've seen, stress can get into a positive feedback loop. The more stressed you are, the more unhappy you become. This unhappiness then results in further stress. A fundamental requirement is to break out of that loop, and a very powerful tool for managing this is laughter.

Spend a few minutes putting together a laughter lifeline pack. Note down everything you can think of that makes you laugh. Not a snide, put-down sort of laugh. In fact, not any nasty or calculated sort of laughter, but sheer, uncontrolled hilarity. It could be certain books, cartoon strips, movies, comedians, TV shows – or just a good evening out with your friends. Once you've got your list, see if you can have one or two of these laughter lifeline elements on call for when you feel down.

We don't find it at all strange to keep first-aid kits on hand in case someone needs some minor physical repairs, so it's rather odd that we don't give any consideration to our mental well-being. Humour and laughter tend to be frowned on in a business context, or when dealing with a serious issue like climate change. Apparently we aren't supposed to enjoy ourselves when we are involved in something important. This sheer madness seems to derive from some Victorian work ethic, or the strange concept that humour and laughter are somehow not professional. Whatever the cause, it needs fighting.

Laughter is a multiple stress-reliever, helping on both the mental and physical levels. There are chemical processes at work in the brain that partly explain this, but a lot of it is down to the benefits of sheer enjoyment. Indulge! This is a particularly good technique to apply when others are suffering from stress. Get them involved in an evening of laughter. A trip to see a top-rate comedian performing is probably best of all. There's something very refreshing about laughing with a group of other people.

Don't forget to bring play into your life as well. Play is a valuable technique that eases stress very naturally. It's sad that we lose a lot of our ability to play as we grow up, when we need it more than ever. This isn't about playing sport – in fact, most sport isn't play in the sense of being fun and unstructured.

Find some form of play in which you can totally lose yourself. It might be playing video games or board games or silly party games. It might be conjuring up a fantasy world on the train, or trying not to step on the cracks in the pavement, or even saying 'boing' every time you pass someone with red hair. Just play.

Such play can be undertaken at pretty well any time of day (especially the types that don't involve technology) and can last a few minutes or hours. The great thing about play is not only are you putting aside all your everyday stressors, but the activity you are involved in is deliberately not important. It doesn't matter what happens, it is just play. We should all indulge in play more often.

Take a break

You are working under pressure. Time is short and there is a huge amount to be done. So you work long into the night, steaming open your eyelids with cups of coffee, hardly stopping. Stress builds and builds as the deadline grows near. Keeping going without a break seems a natural thing to do when under pressure.

Unfortunately there is overwhelming evidence that this is not a great way to get the most out of your brain. The amount of information retained and the quality of your output drops off after spending a long time working at the same task. By taking a series of short breaks, much more can be achieved. There isn't a magic length for the timespan, but most people find working in chunks of between forty-five minutes and an hour, with breaks of around five minutes, will overcome the deterioration.

Sleep

You only have to speak to someone who has had a baby for the first time to realize how stressful going without sleep can be. We're all conscious of the limited time there is to live a life and want to squeeze every last drop out. Doubly so when struggling with the impact of climate change. That's fine, and may well be necessary, but insufficient sleep is a surefire remedy for stress.

All that's required for this exercise is to spend a little while thinking about your sleeping and what you can do about it. If you have regular variations in sleep patterns of more than about an hour a day, you are likely to suffer. Most of us should have around seven hours' sleep a night on average, though the amount does vary significantly from person to person – only you can know what you really need. But it's one thing to prescribe sleep, and it's another to get it.

If you have trouble getting off to sleep, there are mental and physical techniques to help. Make sure you aren't trying to remember something as you settle down. If a task you need to do next day is nagging at you, make a note of it, even if it means getting out of bed. Don't try to go to sleep straight after a passionate discussion – wind down first. If something is going round and round in your head, sit up and pin it down, don't try to force sleep on yourself. When you do lie down, use a calming, tranquil mental image to help you drift into sleep. Some people find a warm, non-stimulating drink helps. Also a warm bath, followed by getting straight into a warm bed, can be effective – but make sure the bath isn't hot, as this will stimulate rather than wind you down. Try relaxation and mental exercises before resorting to drugs. Sleeping pills are rarely an effective answer.

Sleep deprivation piles stress on stress until you are almost driven mad. If you aim for the amount of sleep you need to feel well, rather than the smallest amount of sleep you can get away with, you will underpin all your other efforts in managing stress.

Break the negative cycle

Perception plays an amazingly important part in stress. Our bodies are easily fooled at this gut level. You can generate unnecessary stress – or eradicate it – by modifying your state of mind.

Most of us are pessimistic part of the time; some make a career of it, always finding something to moan about. You can make a huge difference to your stress levels by forcing yourself into Pollyanna mode. Bearing in mind the impact of climate change, you might find it hard to accept there is anything to be cheerful about. Forget that. Being miserable about things you can't influence is a sure way to an early grave. Take on the stuff you *can* change with a will, and get yourself a positive attitude. Easier said than done? It's surprisingly easy.

Spend a few minutes thinking through the ways that you are naturally pessimistic and optimistic. Then, for the next week, be really conscious of your attitude. Put in various places (beside your bed, in the car, on your desk) a little reminder of this exercise – a picture of a half-full glass. Whenever you get a chance to make an observation that could go either way, or even have a thought, force it into the optimistic. Don't look at a cloudy sky in the evening and say 'Looks like rain tomorrow'; instead say 'It may well clear in the night.' Make the glass always half full.

The more you force yourself to act optimistically, the more you will actually feel positive. The more you feel positive, the more you will feel in charge and the less stressed you will be. Usually, the difficulty is maintaining the stance – hence the little visual reminders. After a week you may well find it sufficiently habitual that you do it without being reminded.

Being optimistic comes naturally to most children, but it gets worn away as we get older.

Don't dismiss rituals

Ritual, a regular practice, is a powerful bulwark against stress. Having an established ritual for a small portion of each day provides an anchor for a fast-changing life. This is particularly important with a change-driven stress like the impact of global warming.

In itself, the ritual doesn't have to be big or significant. And there are times when you will have to abandon it with good grace. But the norm should be that your ritual is carried out. Evenings are generally the best time, as the ritual helps you refocus after the workday. It might be having ten minutes with a glass of red wine, or reading a story to the children, or watching your favourite soap opera, or a religious observance – the activity itself is less important than the ritual nature.

Sit down for five minutes and think about your life. What elements are potential rituals? How can you protect them? Try to give yourself a daily ritual, preferably in the evening. You might also like to establish a weekly ritual at the weekend – here the evening setting is less important, but again it needn't take up too much time. Try it for a few weeks to get in the swing.

⚡ POWER TIP

Why rituals aren't as bad as they're painted

Ritual has got a bad name. If you say 'It's a ritual with him', the tone is condescending. The implication is that having a regular pattern of doing things means being stuck in a rut. There's a germ of truth there. If everything you do has to fit an unchanging pattern, then you are doomed in a world of climate change. But it's not that simple. No matter how flexible you are, you can benefit from a small core of ritual. Like the family and the home, it provides stability in an otherwise chaotic environment. Look how important rituals seem to have been in prehistoric society. Rituals mustn't dominate, but there should be a thread of them in your life.

Many of us already have a ritual but don't recognize it, and certainly don't give it the importance it deserves. Others currently lack such an anchor and will benefit even more from this exercise.

Music soothes the savage breast

Often stress strikes at a time when you can't do anything about it. One of the reasons road rage is so common and extreme is that the driver is highly restrained by the physical and mental requirements of the task. You can't start exercising or have a massage to overcome stress in such circumstances – but you can use music. The right sort of music will lower your heart rate, get you thinking in a more relaxed way and generally put your stress into perspective.

Note 'the right sort of music'. Not all music is de-stressing. Anything with a fast beat and a heavy, pulsing bass will act more as an adrenaline booster than a relaxant. And don't think just because a piece is classical that it's calming – there's plenty of classical music that will push up your heart rate.

Look for slow, calm music, reminiscent of flowing water and happy, untroubled times. (I have seen Barber's Adagio for Strings, or the Agnus Dei, his incredibly evocative vocal setting of the same music, cited as an example of soothing music. It's certainly slow, but has such a deep-seated core of despair and longing that it isn't the right choice here.) The music can be anything from classical to folk as long as it has the right effect. I find Tudor and Elizabethan church music, which combines a steady, flowing quality with spiritual depth, particularly effective. Try a few different styles and see which suits you best.

Make sure that the music that is relieving your stress isn't stressing others. The tinny rattle of overheard MP3 headphones or the booming bass of a car passing with its stereo up too loud causes plenty of irritation for others. Try to keep your music to yourself. With appropriate music you can distance yourself from stress and put it into perspective. There's nothing better if you are stuck in a traffic jam, or just in need of relaxation and stress relief wherever you are.

Cope better with time

There's a fallacy that you can manage time. Time can't be managed, it just grinds along, second by second. But you can manage how you make use of it and how you respond to time pressures, which otherwise can have a strong stressing impact.

Perhaps the worst time-related stress is the result of bad scheduling. Get things wrong and you are constantly late, constantly being nagged by others. One simple way of improving things is to build buffers into your timings. Build in some extra time for things to go wrong. Have something you can do with you, or a diversion planned – that way, if everything runs on schedule and you've time to spare, you have the gift of some free time, rather than time to waste.

If things go so wrong that even your buffers won't help, try to be resigned. The world won't end because you are late. Make sure you have a mobile phone and contact numbers for wherever or whatever you are going to be late for. You have done all you can – now forget about it until you can next take action, and do something positive.

The other key to time management is breaking your goals – the things you want to achieve – down into tasks, and breaking up those tasks into small enough chunks to be able to handle in a day or less. See page 236 for more details.

POWER TIP
Top-ten list

A powerful way to keep on top of your time is to make a top-ten list each week. This should contain the ten most important things you want to achieve in that week. Keep the list somewhere easily visible. When you get a call on your time, glance at the list. Will this activity help towards one of your top ten? If not, it doesn't mean you should always say no, but it gives you the chance to consider whether or not to go ahead.

If you find items roll on from week to week, staying on your top-ten list, they either aren't important, and should be dropped from the list, or they are too big to complete in the next week – break them down into smaller chunks.

Relaxation by breathing

It's a self-evident truth that breathing is a good thing – but there's breathing and there's breathing. Firstly, as all singers know, there are two types of breathing: with the chest muscles and with the diaphragm. The latter is more controlled and gives a much deeper breath, yet it tends to be underused, particularly by those under stress.

First try to feel that diaphragmatic breathing. Stand up, straight but not tense. Take a deep breath and hold it for a second. Your chest will rise. Now try to keep your chest in the 'up' position while breathing in and out. You should feel a tensing and relaxing around the stomach area. Rest a hand gently on your stomach to feel it in action.

Once you've identified that breathing from the diaphragm, lie on the floor or sit comfortably in a chair. Close your eyes. Begin to breathe regularly: count up to five (in your head!) as you breathe in through your nose. Hold it for a second, then breathe out through your mouth, again counting to five. Rest a hand on your stomach. Don't consciously force your ribcage to stay up now, but concentrate on movement of the diaphragm. Your stomach should gently rise as you breathe in and fall as you breathe out. Feel the stress flowing away.

POWER TIP
You can breathe anywhere

One of the great things about breathing exercises is that they can be performed pretty well anywhere. For instance, although driving a car isn't the ideal position, you can still indulge in deep breathing. A regular five-minute session of breathing properly will provide the foundation for many other stress-management techniques. It is simple and very effective.

Relaxation by numbers

If the impact of global warming is getting on top of you, try a little systematic relaxation. It needn't take long, but you do need somewhere quiet to be able to either lie down or sit in a very comfortable chair. Close your eyes, lie back and relax. Try to clear your mind of all thoughts.

Now focus your attention on the parts of your body, working from your head down to your toes. As you consider each section, tense and relax the muscles a few times, holding them tense for a couple of seconds, then relaxing to a long, slow breath. Try to keep your concentration on the area you are exercising – don't let your thoughts drift off to problems or concerns.

When you have worked down the body, lie still, breathing slowly, keeping as much as possible to a mind blank of thought for another minute or so. While doing so, keep your muscles as relaxed as you can. When you have finished the exercise, don't jump up, but gently open your eyes and stand slowly.

Although this technique requires a haven from the stressful world, it can be carried out quite quickly, and is a good emergency defence when things are getting on top of you. You can combine this exercise with a breathing exercise like that on page 227.

Relaxation from the inside

For some, meditation is a natural part of life, for others a symptom of the lunatic fringe. In fact, there is nothing extreme or 'New Age' about meditation, nor does it require acceptance of a particular philosophy. Find somewhere where you can sit quietly and comfortably – unless you are very supple, cross-legged positions should be avoided. Reduce sensory distractions to a minimum. Breathe slowly and evenly. Imagine that everything is slowing down.

Then, find a focus. This can be a meaningless set of syllables, or a simple phrase, or a very calm image like a great, unruffled lake, or a single leaf. To begin with, you may find it helps to have a physical object to provide the focus, but before long you will be able to do without it. If you use an object, don't think about any associations or properties it has. Keep your focus on the entire object. For a few minutes (with your eyes closed unless you are using something physical), let your focus fill your mind. The stress will drain away – but don't think about stress, or its causes.

Initially you will find it hard to keep focused. Your mind will wander. When you notice this, bring yourself back. Don't go too long to start with. Begin with a few minutes and build up to maybe fifteen minutes. Most of the world's religions from Christianity to Zen Buddhism – and plenty of non-religious groups – practise meditation. You can see it as an opportunity to explore inner spirituality or as simple mental/physical exercise. Whatever your view, it works.

Meditation is hard to do without a quiet space and a few undisturbed minutes, but its powerful impact on stress levels makes it well worth trying. It can be helpful to use a timer to avoid worrying about how long you have been meditating, particularly if you have another activity to undertake later.

Learning to love change

Change can be the bane of our life, or the only thing that makes life worth living. Everyone has a change continuum from the level they enjoy to the level that makes them highly stressed. The pace of change induced by global warming is such that most people will be pushed well outside their comfort zones. This is why knowing how to deal with stress is such an essential component of the survival kit.

Spend a few minutes building a change map; to do this, you will have to think about how you resist change-based stress. There are two primary weapons involved in coping with change without being stressed. The first is your anchors. What do you have to return to that remains constant? Write your anchors on the map with circles round them. They could be your family, your friends (or one special friend), your home, your religion, your pet, your rituals – the things you turn to when everything else is in turmoil.

The second weapon is learning to love change. Not espousing change for change's sake, but taking a particular change, understanding the benefits it can bring and making a conscious effort to buy into it. This is possible much more than we normally allow. Draw the prime areas of change on your map, highlighting those where the change is distressing. For those change elements, note down what's good about them. Tie them back, where relevant, to your anchors. Try to focus on those positives rather than your negative feelings. Of course, you probably wouldn't want the changes that global warming has forced on you, given a choice. But now it has happened, it's better to get on top of the change and accept it than to try to deny it ever happened.

Resistance to change isn't inherently bad, but where it's a change you can't directly influence, like climate change, getting worked up about it is futile. Change is ever-present and particularly so at the moment. It is always stressful. Learning to cope with it is a major stress-relief skill.

Benefiting from a book

Under the right circumstances, reading a book is a very calming activity. This isn't a prescription to deal with peak stress. If you are extremely worried about something, or bursting for action, you will not be able to get into a book. But books are ideal for chronic stress, when the little things in life (and climate change) wear you down.

Most of us don't read enough – in breadth or quantity. Find two slots a day to do some reading. Then look at your choice of books. You need something that will take you away from everyday pressures. Don't go for a 'quality' novel about depressing people and their agonizing lives. The book doesn't have to be upbeat, but the last thing you want is to be depressed. Often genre fiction can be effective. After all, a fantasy or a murder is unlikely to reflect your everyday problems. Equally, readable non-fiction can work well. Look at areas like travel, biographies, popular science or history.

There are many reasons for reading. Stress-management is only one component. Sometimes, perhaps standing up on a crowded commuter train, reading passes the time without really doing anything about stress. To get the best stress-relief you ought to be sitting in a comfortable chair with no disturbances. Just because this technique is applicable only to the everyday accumulation of small stresses does not mean that it is trivial. Keep up that reading.

POWER TIP
Don't be a media snob

Books aren't the only medium that can be effective. Don't dismiss TV and movies because they're downmarket. Similarly, computer games can be good for stress-relief. Adventure games have a similar effect to a novel, while an action game might push up the adrenaline levels temporarily, but will be cathartic in taking out your stress on a clear, identifiable enemy.

Regaining traditional entertainment

Storytelling

If climate change results in the collapse of many of our modern forms of entertainment, or disrupts the electrical power that makes TV and radio usable, it's important that we relearn some of the traditional ways to keep ourselves entertained. Storytelling goes back as far as language, yet though it's still a given that young children love to hear stories, it may surprise you how effective good storytelling can be for an older audience too.

Even a group of cynical teenagers can, with the right atmosphere, fall under the spell of storytelling. On a dark evening, perhaps around a fire, a collection of spooky ghost stories can be effective for any age. Part of the essence of making modern storytelling work is to recognize that some of the old tricks aren't relevant any more. Some storytelling traditions rely heavily on repetition – this works only with very young children, and can quickly irritate a more sophisticated modern audience. Instead, keep the story brisk and rapidly twisting and turning. Don't get too poetic, go for the gut. With a difficult audience of teenagers, throw in some shock tactics.

The ideal is to get everyone in a group telling a story in their turn. They don't have all to be epic sagas – a good joke will do – but the effect can be spiritually warming, bring the group together and emphasize our humanity in the face of frightening circumstances.

When making up a story for a younger audience, involve the children in the story as characters, and throw in plenty of fantasy, animals and any subject you know is particularly of interest. Put your characters under threat. They should be in danger of their lives or worse. Give yourself a challenge – push the characters to a cliffhanger, then find a way out. With any luck, you will enjoy the process as much as your children.

Making music

Music is one of humanity's greatest gifts. We have become used to having our music packaged and presented through electronics. If climate change brings interruption of power supplies and difficulty in getting entertainment through the box in the corner of the room, we need to look again at the joys of making music. Most of us aren't musical superstars, but we can hold a tune, sing a song, maybe pick out a tune on an instrument.

The biggest hurdle to be overcome is our expectations – we have become used to the unblemished nature of modern commercial recordings. This is music that has been massaged and modified until it has such an unnatural air of perfection that ordinary performances seem weak. It takes time, but you will need to get back to appreciating this raw sound, rather than processed perfection.

Dig out the cards

Amazingly, even working as hard as they had to in the olden days, people had some time for entertainment. Without TV or movies, games resurface in importance, whether a simple game of cards or a sophisticated board game. If climate change makes normal twenty-first-century entertainment difficult, don't collapse in misery: go back and find the entertainment that games have to offer.

If you don't have many games, or you are fed up with the ones you have, it can be surprisingly easy to improvise something yourself, particularly if the requirement is to entertain children. Try putting together a treasure hunt around your house, or outside if it's appropriate. Compile a series of cryptic clues that lead from place to place, each guiding the players to the next one until you finally reach a prize. A game like this can be very satisfying, both for those taking part and the person who constructs it.

Books as the new currency

Take a look around your house. Do you own plenty of books? It's a long shot, but building up your home library could be a recipe for future wealth in a world radically changed by global warming. In the future, you may be more reliant on what you can exchange than you are on money. If the shops aren't open, then money becomes a doubtful currency. It's much better if you can barter, exchanging one commodity for another. Your expensive electronics aren't going to be a lot of use if there's no power, but books have the advantage of increasing in value in difficult times.

Not only is a book a handy unit of barter, books are unusual in being still usable without power when other forms of entertainment are unavailable. And the more books you own, the better your chance of having something useful to exchange. It might be time to add to the books on your shelves, while you can.

Problem-solving for beginners

Everyone has problems in their lives, but for those of us who have a job and a home, many of the basic problems of survival are taken out of our hands. We don't need to worry about how to find food; we just head off to the supermarket. Without the safety blanket of our technological civilization we face many more problems. Even if climate change doesn't totally shut down our technology for a time, the issues it raises will mean that we face problems more frequently and those problems may well be new and challenging.

Problem-solving and generating new ideas are skills that can be learned and practised like any other. The trouble is, few of us are taught these skills in school. Yet just a little guidance can make a lot of difference to our capabilities. Here's a short, four-stage approach to improve basic problem-solving.

The first essential is taking the time to think. A very simple technique that can be very effective in problem-solving is to go out for a ten-minute walk. Don't explicitly look for solutions, but let the problem filter through your mind. Have a pencil and paper, or some other way of making a note like a voice recorder with you. Jot down any thoughts you have. It's amazing how often you can come up with an idea using this process, particularly if the walk is in the countryside or somewhere else with little human distraction.

Next, make sure you are dealing with the right problem. All too often we can rush in and try to solve a problem, only to realize too late that we're attacking the wrong issue. Think through what you want to do in solving the problem. Why is there a problem? What are you aiming for? What stops you from getting from where you are now to your goals? See if there's something different you could do – for example removing those obstacles – which will solve your problem more effectively.

Thirdly, see what's suggested by your resources. What have you got available? How could you use it? Think outside the normal uses. Coins, for instance, have been used as wedges to prop up furniture, as impromptu screwdrivers and much more, not just as a form of currency.

Finally, give your assumptions a real hammering. It's assumptions that usually stop us from being creative and coming up with good ideas. Assumptions about 'how we do things round here', assumptions about how the world is, and assumptions about what is possible. Almost every assumption can be removed, at least temporarily. You might have to reinstate the assumption because, for example, to do otherwise would break the law (or a law of nature), but there's nothing to stop you from ignoring the assumption temporarily while you come up with a great idea, then reinstating it and modifying the idea to cope.

⚡ POWER TIP

Don't try to take off wheel nuts with your bare hands

When we're under pressure it's easy to forget the importance of taking time to think. We fall into the trap illustrated by that old chestnut: 'When you're up to your neck in alligators, you haven't time to think about draining the swamp.' But the swamp is an illusion. I'd like to offer you a much more realistic metaphor for those who feel they haven't time to give some thought to coming up with a creative idea.

Imagine you are driving along in a car, and see a friend broken down at the side of the road. You stop to help. The friend is kneeling by one of the front wheels and trying to take the wheel nuts off with his bare hands. 'Hold on,' you say, 'don't be silly. Why don't you go and get a wrench from the car?' The friend shakes his head. 'I'm in a real hurry,' he says. 'I don't have time to go and fetch the wrench.' So he stays there, forever trying to get those wheel nuts off with his bare hands.

Problem-solving and coming up with creative ideas is like this. If you say you don't have time to stop and think, you don't have time to use a technique or to consider your options, you just have to get the problem solved, then you're like the guy by the side of the road, trying to get those wheel nuts off. Of course there will occasionally be situations where there genuinely isn't time to think and you have to make a decision and go with it – but most often you will reap huge benefit from taking the time a good decision needs.

Breaking it down

Some of the problems faced by those trying to survive in a rapidly changing world can be meaty ones that aren't going to be sorted out with a quick fix.

When you are faced with something that can't be sorted out in an hour or two, make sure you are using the tried and tested approach that businesses employ in dealing with bigger projects. Break the solving of your problem down into a number of chunks. Typically it's best to make these ones you can work through in forty-five to ninety minutes. After a period of that length, the brain is looking for a change. Give yourself a couple of minutes' break doing something completely different. It doesn't have to be a rest, but does have to be different.

By breaking down your task, whatever it happens to be,

into a series of bite-sized chunks you can make it much more likely that you will succeed. The only problem with this approach is that you do need to manage what you are doing or you can end up with a whole pile of projects that are started but not finished. Don't just break your job up into pieces, but set yourself a target for when you are going to complete each piece. Keep these targets updated regularly. Targets shouldn't be an end in their own right, but they will help you to finish what you began.

Getting the right environment to be creative

Everyone is creative. Admittedly some find it easier than others, but everyone can have good ideas. Yet put even the best thinker into the wrong environment and they will struggle to come up with anything. In part this is a matter of stress. The advice earlier in this chapter on managing stress isn't just so you can feel more comfortable and be healthier. It's only by reducing stress that we can come up with good ideas and solve problems effectively. It's hard to be creative when, for example, someone is leaning over you, demanding an idea.

This is because of the way the brain uses thicker, well-used neural connections when under pressure. These will be the well-trodden paths of thought, rather than new ways of thinking.

The other aspect of environment is getting away from distraction. This doesn't mean that you have to be suspended in a sensory deprivation tank, but all the paraphernalia of modern life, from mobile phone to MP3 player, don't help when you are trying to think. Even if there are problems with the technology thanks to climate change, there can be plenty of distractions. Getting away to think makes all the difference. It can help if you engage in some repetitive activity, whether it's walking or digging the vegetable plot – this helps the mind settle into the daydream state which is best for new ideas to form in and problems to be solved in.

Generating ideas the random picture way

There are many techniques to help you to reliably solve problems with creative flair and to generate ideas – see page 263 for sources of further information. Here I will just illustrate a handful that are especially powerful.

The first involves using a randomly selected picture. You need a picture with plenty of interesting elements in it, but it shouldn't have anything particularly to do with your problem. The most common way to select these pictures is to use the image search at www.google.com or a similar search engine, but if you have to employ the technique without access to the internet, you can use a newspaper or magazine, your old photographs or an illustrated book. Just find a picture with plenty of interesting detail.

Now put your problem to one side for a moment and take a good look at the picture. Jot down what the image makes you think of as a series of key words. What does it remind you of? Perhaps there is some memory it triggers. Look at different parts of the picture, and the image as a whole. You should easily be able to come up with between twenty and thirty associations from a good picture, though seven or eight would be enough.

Now take each of your associations and use it to try to come up with an idea to help with your problem. (You've finished with the picture now – it was just a way to produce these associations.) Don't rush. Take a little time with each association and put it alongside your problem. For instance, say my problem is how to preserve food. I happen to have found a picture of a dog jumping through a hoop in a stage show. One of the things the dog made me think of was a husky dog, pulling a sledge across the snow. So I could look for naturally cold places to help preserve my food. I also, for obscure reasons, thought of the video-game character Lara Croft. An adventurer like Lara carries dried rations. How could I dry my food to preserve it? The dog in the picture has its hair standing on end, as if it has been zapped with static electricity. Could I use an electric discharge to make the food last longer? ... and so on.

The ideas in that example weren't great – you could do much better – but it illustrates how the technique works. It gets you to look at your problem in a different way, because the associations with the picture produce new starting points in dealing with your problem. This is a very reliable technique that almost always generates a good range of ideas you might not think of otherwise.

Taking someone else's viewpoint

Very often it can help to get someone else to take a look at your problem. Seeing something through fresh eyes can produce a whole new range of options. Ask other people, if you can. But you can do more than this. You can force yourself to take a very different viewpoint by imagining how someone else would deal with your problem, someone very different from yourself. You can pick someone you know or know of – work out who would take the most different approach. Or you can use a randomly selected person to help. In the checklist on page 250 you will find a list of 60 different characters.

Pick yourself one of the characters (you could use the second hand of a watch to pick an item at random from the list). Don't switch to another if you think it's not the right person – your only excuse for switching is if you've no idea at all who this person is. Spend a minute or two imagining what it would be like to be them. Then imagine that you, as that character, have to deal with your problem. List the different ways you would deal with the problem as the character. Then back in your normal persona, take those ideas and turn them into something practical you can do.

Inspiration from nature

A powerful technique for creativity is to look to nature for answers. This doesn't mean having the dogmatic idea that 'natural is best', but rather looks to the solutions nature has come up with to similar problems. So, for example, if your problem is to do with building something, look at how a tree is built up, or the exoskeleton of a beetle. If your problem is about dividing things up, look at how ants operate, or how a cuckoo lives.

An obvious reaction is 'I don't want to be like an ant or a cuckoo', but this misses the point. The technique is not about copying nature but about being inspired by it. Use what you see as a starting point, but go far beyond it. In the cuckoo example, for instance, I might see that the cuckoo tries to make the host bird look after its egg without realizing it. So if the problem is how to divide a scarce resource, I might look at how to get people to make the division without realizing they are doing it. Or I might be inspired by the way the cuckoo gives its share (i.e. the egg) to someone else – and at the same time gives the accompanying work of looking after the young bird. Could we allow unfair shares, if someone is willing to take on an extra burden?

When making use of nature as an inspiration it's not necessary to find an obvious parallel like the ones above. In the division problem, one might find the germs of just as many creative ideas in studying a leaf. Pick up a leaf and look at it closely. Examine the surface. Look at the structure beneath by holding it up to the light. We don't normally look at individual leaves, but there's a lot of detail in there. The thoughts that come to us while examining a part of nature in this sort of detail can be equally as valuable a kicking-off point for new ideas.

Creating from scratch

Creativity isn't purely a matter of problem-solving. Sometimes ideas can be more ones to inspire others than a solution to a problem in their own right – the sort of creativity that is often called art. Bearing in mind that everyone is creative in their own way, it might be that you make use of some spare time to be creative for others. Have a go at putting a song together. Play around with paint or sculpting materials. Dream up a poem. This might seem childish, but the fact that it does so is more a sad reflection of the way we have allowed the creative arts to be dominated by children's work and that of professionals. Arguably that's why there's so much art that practically no one is interested in. We have allowed the 'experts' to take it away from us.

If this chapter is about finding the human side to surviving climate change, surely there is a big opportunity here. We can capture art back from 'them' and make it something of the people again. The effortless entertainment provided by modern technology makes us lazy when it comes to art. While the disruption caused by climate change is not desirable, we should enthusiastically grab the opportunity it gives us to all be more creative.

There's a book in all of us

There's an old saying – everyone has a book in them – which is usually extended in the publishing business to 'Everyone has a book in them, and in most cases it ought to stay there.' Most of us, at some time or another, have felt the urge to write a book. The cynicism of the publishers' version of the saying arises from the fact that most of us aren't very good at writing books – but that doesn't stop it from being, for some, a very fulfilling task, even if their work is never read by anyone other than friends and relations. If you have some spare time thrown up in the dark, power-free evenings, all you need is a light and a pen and paper: many great books were written without a computer.

If you have children and like telling them stories, why not try starting with a children's book? This has the advantage of being relatively short, and your children will provide a natural audience. But don't feel this is the only possibility.

Take a few minutes to jot down what really interests you. A possible subject for a non-fiction book might be a hobby, or a subject that gets you really excited. It could be your own life (though bear in mind, if you want this to be a book and not just a diary, the content should be interesting enough that others want to read it). Equally, think about the kind of fiction you can't put down. What really keeps you reading?

With topics in mind, now give a little thought to who might read your book. It may be you will just keep it for your own use or share it with your family. It could be that your book becomes a word-of-mouth hit, even if it isn't actually for sale. Equally it could be that publishing hasn't stopped, or restarts after you've written your book, and it could become the real thing. It's important to be honest with yourself at this stage. You may be utterly fascinated by the mating dances of the lesser crested wombler – but you might find few others who want to read a whole book on the subject. Try to see your book through other people's eyes.

When you've hit on something you really want to write and feel that others might want to read, you've got something to aim for. Don't expect to get it all finished in a week. Part of the positive aspect of writing a book is that it's a big goal, something to aim for and to manage a chunk at a time. Start by planning it out. If it's fiction, try to get a feel for the outline of the plot first, and who the main characters are. Fill out some detail about them for your own benefit, so you know them better as you write about them.

If you are writing non-fiction, try to write a chapter-by-chapter summary – a paragraph or two on each chapter, describing what's in it, and how the reader is going to be carried through. This is useful to help you as you go, and will also help you distinguish between an idea that might work as a short article, and one that can be sustained for a whole book.

Then go for it. Try to write several times a week – it need only be for half an hour at a time. Find yourself somewhere quiet to do it. If electricity is available so that you can write on a computer, keep track of your word count to see how you are getting on. If you are writing on paper, use page count as a guide. Most of all, enjoy it. This is an exercise in stretching yourself, but if writing becomes a chore, give it a break.

Resources

Pen and paper

There are few resources for making the most of your human capabilities that are more versatile, flexible and reliable than pen (or pencil) and paper. There is no power requirement, so you don't have to worry about the impact of a blackout. Nothing is lost to power surges. Whether you are playing a game, writing a book or coming up with good ideas, you can't go far wrong with paper and pen. A large-format notebook (A4 or letter size) and a few pens and pencils provide a basic minimum.

Games

Stockpile a few games. Make sure some are suitable for playing with a wide range of ages. Board games are great – have some of these too – but there are relatively few that suit all ages, especially very young players. Include several packs of cards (and maybe a book of card games if you aren't too familiar with them) and simple games like Connect 4 or Kerplunk. Word-based games like Boggle and

Scrabble can be very effective too. A stash of games is very useful if you have to entertain yourselves, and they will make good bartering objects.

Musical instruments

Making your own music is a great fallback, both on your own and with friends and family. Make sure you have some cheap simple instruments like a recorder, along with a teach-yourself book if you can't already play. Non-powered keyboards like a piano are great too, or if your budget won't run to this, a battery-powered keyboard can be effective for picking out tunes. Don't worry if no one plays at the moment – learning to play can be part of the challenge.

Pet solution

There is ample evidence that pets reduce stress levels and anxiety. In some scenarios arising from global warming, keeping a pet could be its own source of stress. If there just isn't enough food to go round, you may at some point have to decide between the pet and people. But assuming things aren't that bad, pets have a lot going for them. The positive effect of pets on stress is now recognized to the extent that pets are allowed into some hospitals to help the recovery of patients. Take some time to think about the lifetime of the pet. You may be taking on a commitment that runs into decades. Consider how much maintenance is required. Is no one at home all day? Are you away a lot? How would you cope?

The actual technique once you've got the pet is to spend some quality time with it. Stroke the pet if feasible. The sessions needn't be long to be noticeably beneficial.

There is a rough spectrum of pets: the higher up the scale, the greater the commitment, but (usually) the greater the return. A few key points:

❐ Goldfish – very low commitment. Grow much bigger and last much longer in a pond; watching them in the pond is more therapeutic than gazing at them in a tank (though weather-dependent).

❏ Tortoise/turtle – can be very long-lived, and owners will tell you they have great character, but not exactly cuddly. Relatively low maintenance.

❏ Hamster – only lives around two years, but can be stroked. Very variable in personality. Sleeps a lot in the day.

❏ Rat/guinea pig/rabbit – longer life than a hamster; more commitment. Select a rat for intelligence, a guinea pig for docile stroking and a rabbit . . . if you like rabbits (they have the advantage that they can be house-trained).

❏ Cat – looks after itself a fair amount of the day. Less docile than a guinea pig and not as friendly as a dog, but plenty of personality.

❏ Dog – the ultimate stress-relieving pet, but also by far the biggest commitment. Larger dogs can also provide some security in dangerous times.

If owning a pet isn't practical, see if you can gain regular access to someone else's pet.

Fallout shelters

It's possible that some of the more extreme impacts of global warming could result in the need to build yourself the stress equivalent of a fallout shelter.

Spend around five minutes thinking about the personal elements of life that help you deal with stress. Imagine you were building yourself a stress-proof bunker (not a physical bunker, but an emotional one). What would you incorporate? It might involve quite mundane items. Typical contents might be:

❏ Family photographs

❏ A print of your favourite painting

❏ Photographs of restful scenes

❏ A pair of slippers

❏ A comfy chair

❏ A sound system

. . . but it might involve something more bizarre.

Now get together a list of small actions that would be needed to assemble your anti-stress shelter. Make sure that some of these actions are carried out in the first week. The image of a bunker or shelter might be misleading. You aren't trying to cut yourself off from the world or all its influences. It is stress and its fallout that you are sheltering from. The stress shelter isn't going to keep all stress away from you, but it will provide excellent protection against the worries that global warming and the world can throw at you.

CHECKLISTS

Getting enough sleep

Are you getting enough sleep? It's an essential if you are going to win through a stressful situation and have enough energy to deal with anything that climate change can throw at you. Most adults require around seven hours a night, but the amount varies. If you find that more than one or two of the checks below apply to you, see if you can add to your sleep time. Try going to bed half an hour earlier. Cut out or reduce reading in bed. If this still isn't enough, keep adding half-hours until you feel better.

Difficulty getting out of bed .... ❏

Tendency to go back to sleep before you have a chance to get moving .... ❏

You fall asleep as soon as you sit down after a day's work .... ❏

Droopy eyes and a tendency to nod off while driving ❏

You feel hung over in the morning, even though you weren't drinking .... ❏

Physical checks

Just how stressed are you? This is the first of two checklists to monitor the types of reaction that stress typically induces. Which, if any, of these physical symptoms are you subject to?

Regular indigestion .... ❏

Inability to sleep well .... ❏

Aches and pains that respond to massage .... ❏

Eczema, spots and other skin complaints .... ❏

Frequent headaches .... ❏

Difficulty catching your breath .... ❏

Feeling dizzy or shaky .... ❏

Breaking out in a cold sweat .... ❏

Tingling in your palms .... ❏

Note that not all physical symptoms of these kinds are caused by stress. If you have symptoms that continue, check with your doctor. If you have more than one symptom from this and the next checklist combined, make sure that you work on your stress-relief as soon as possible.

Emotional checks

Stress reactions aren't limited to the physical – after all stress is very much a mind/body problem. This second checklist looks at typical stress symptoms in the emotional and mental areas. Which, if any, of these symptoms are you subject to?

Forgetting things a lot .... ❑

Decision-making is difficult .... ❑

Your driving has deteriorated .... ❑

You feel restless .... ❑

You get frustrated with others .... ❑

Unusual impatience .... ❑

Mood swings .... ❑

Lack of concentration .... ❑

Everything seems pointless .... ❑

You can't keep on top of things .... ❑

You feel defensive .... ❑

As with physical symptoms, bear in mind that many of these reactions can be due to illness as well as to pure stress. If you have symptoms that continue, check with your doctor. If you have more than one symptom from this and the previous checklist combined, make sure that you work on your stress-relief as soon as possible.

Depression checks

Depression is more than just a matter of feeling miserable, it is a clinical condition. Confusingly, depression comes in two forms: it can be driven by outside events, or can come from within. The latter case (endogenous depression), where there seems to be no obvious cause, is a purely medical condition requiring professional attention.

Look out for some of these possible indicators of depression:

Always feeling tired, however much sleep you get .... ❑

Feeling worthless and without value .... ❑

Significant loss of appetite .... ❑

Problems managing at work, where previously
this wasn't the case .... ❑

Always feeling sad, or simply blank .... ❑

Unusual consideration of, or obsession with, death .. ❑

Frequent anxiety without any particular cause .... ❑

Lack of interest in your social life, or sex .... ❑

Can't put two thoughts together .... ❑

Although the stress-management techniques in the 'Solutions' section may help to relieve depression that is driven by outside events, it can't help the internal form. If you feel stressed make sure that depression isn't part of the problem – if in doubt, and the symptoms persist, contact your doctor.

In many instances depression *is* simply a reaction to stress, but it is important to recognize the possibility that this might not be the case, so that alternative action can be taken where appropriate.

Character list

A list of sixty different characters to use in the creativity technique on page 239. The simplest way to pick one is to check the position of the second hand of a watch and use the character given the corresponding number in the list.

1. Marilyn Monroe
2. Queen Elizabeth I of England
3. A casino croupier
4. Adolf Hitler
5. Attila the Hun
6. A librarian
7. William Shakespeare
8. Macbeth
9. A Roman centurion
10. A plumber
11. Dr Livingstone
12. David Beckham
13. Eeyore (from *Winnie-the-Pooh*)
14. The current US president
15. A pet rabbit
16. A mass murderer
17. A prison warder
18. An action-hero movie star (pick one you know: John Wayne, Bruce Willis, Arnold Schwarzenegger etc.)
19. Zorro
20. Robin Hood
21. A computer programmer
22. Snoopy
23. Ludwig van Beethoven
24. Karl Marx
25. Groucho Marx
26. A spin doctor
27. The current prime minister of the UK
28. George Washington
29. A nasturtium
30. A surgeon

31. The head teacher of a large inner-city school
32. A circus clown
33. A circus trapeze artist
34. A poet
35. A Roman Catholic priest
36. A rabbi
37. Confucius
38. A great architect (choose a real one if you know one)
39. Sherlock Holmes
40. Hercule Poirot
41. A Martian
42. Winston Churchill
43. Count Dracula
44. Mao Tse-Tung
45. Billy the Kid
46. Tutankhamun
47. Charles Darwin
48. An ant
49. A court jester
50. Orville Wright
51. Paul McCartney
52. An astronaut
53. A blind person
54. A deaf person
55. Mr Spock
56. Donald Duck
57. A New York cab driver
58. The Pope
59. Superman
60. A beggar in Bombay

APPENDIX

GWSK RESOURCES

First point of call – www.gwsk.info

The Global Warming Survival Kit website, www.gwsk.info, contains the most up-to-date versions of all the links and information below, plus a wide range of extra material to support your global warming survival.

This appendix lists some recommended books and a selection of relevant websites. It is not an endorsement of any of the sites mentioned.

Teetering on the Brink

Further reading on and links to the latest information on climate change and its impact on our lives.

Read

The Atlas of Climate Change, Kirstin Dow and Thomas E. Downing (Earthscan, 2006)
An atlas focusing on climate change, its impact and the warning signals we face.

Climate Change Begins At Home, Dave Reay (Macmillan Science, 2005)
After a brisk introduction to the realities of climate change provides practical guidance on doing your bit to put it off.

The Chilling Stars, Henrik Svensmark and Nigel Calder (Icon, 2007)
Although there is consensus that human intervention has influenced climate change, there can be other influences as well: this book describes the potential for cosmic rays and the Sun to contribute directly.

Field Notes from a Catastrophe, Elizabeth Kolbert (Bloomsbury, 2006)
An arresting exploration of the impact of global warming using specific examples around the world.

Global Warming: The Last Chance for Change, Paul Brown (A. & C. Black, 2006)
An easy-reading assessment of global warming and its impact. Doesn't pull its punches, though the DIY solutions may be a little optimistic.

Heat, George Monbiot (Allen Lane, 2006)
Focusing on the warming itself, a very readable assessment of what can be done to stop the planet from warming up.

An Inconvenient Truth, Al Gore (Bloomsbury, 2006)
High-profile book, putting the case for the reality of climate change and the possible solutions.

The Last Generation, Fred Pearce (Eden Project Books, 2006)
Assesses our understanding of climate change from the early pioneers to the latest thinking, and includes a clear explanation of one of the least understood and most important aspects of climate change, the influence of positive feedback and tipping points.

The Rough Guide to Climate Change, Robert Henson (Rough Guides, 2006)
A surprisingly readable and scientifically well-informed summary of the nature of climate change and its impact.

The Weather Makers, Tim Flannery (Allen Lane, 2006)
Good general introduction to the history and impact of climate change.

When the Rivers Run Dry, Fred Pearce (Eden Project Books, 2006)
A fascinating and disturbing combination of travel book and exploration of the increasing incidence of drought around the world.

Browse

Australian Climate Change (www.bom.gov.au/climate/change)
Australia's Bureau of Meteorology gives the facts and figures for the Australian region in an approachable site.

Canadian Climate Change (www.climatechange.gc.ca)
The Canadian government's climate-change site.

Climate Challenge (www.climatechallenge.gov.uk)
The UK government's climate-change site: relentlessly populist, but plenty of content.

Climate Change from the BBC Weather Centre
(www.bbc.co.uk/climate)
A glossy but useful introduction to the basics of climate change.

David Suzuki Foundation (www.davidsuzuki.org)
A Canadian campaigning organization centred on sustainability but with a considerable amount of information on climate change.

Dynamic maps of sea-level rise (www.geo.arizona.edu/dgesl)
Pick any location and see how sea level changes.

EPA Climate Change (www.epa.gov/climatechange)
The US Environmental Protection Agency's take on global warming: includes an effective kids' site.

Global warming: early warning signs (www.climatehotmap.org)
World map showing hotspots for different types of warning signs of the impact of climate change.

Hadleigh Centre information (www.metoffice.gov.uk/research/ hadleycentre)
A wide range of information from the UK government's organization specializing in climate change.

Intergovernmental Panel on Climate Change (www.ipcc.ch)
The world's organization tasked with studying climate change. Rather heavy going, but definitive.

Royal Society on Climate Change (www.royalsoc.ac.uk/ climatechange)
The UK's foremost scientific organization with an effective climate-change site, especially the 'facts and fictions' section exploring misleading arguments.

SciDev on climate change (www.scidev.net/climate)
A fresh look at the science of climate change from a developing-world perspective.

Union of Concerned Scientists (www.ucsusa.org/global_warming)
Science-based US pressure group.

Powerless

Advice and technology to help cope with blackouts and other losses of power.

Read

Home Heating with Wood, Chris Laughton (Centre for Alternative Technology Publications, 2006)
Practical, comprehensive guide to using wood as a fuel (plus combination wood/solar heating).

The Homeowner's Guide to Renewable Energy, Daniel D. Chiras

(New Society Publishers, 2006)
Looks at alternative sources for energy in the home and how to move to independence from the grid. A little optimistic about the practicalities, but plenty to think about.

How to Live Off-grid, Nick Rosen (Doubleday, 2007)
Combines the author's experience of attempting to live off grid in a camper van (though he does make frequent use of the mobile phone 'grid') with a guide to the benefits and pitfalls of stepping away from the grid.

How to Live Without Electricity and Like It, Anita Evangelista (Breakout Productions, 2002)
A touch simplistic, but covers a wide range of non-electric approaches from solar ovens to wind power.

Natural Home Heating, Greg Pahl (Chelsea Green Publishing, 2003)
Concentrates on the options for keeping your home warm using direct renewable energy sources.

Off the Grid: Modern Homes and Alternative Energy, Lori Ryker (Gibbs M. Smith, 2005)
An explanation of how to live unconnected to the grid for both electricity and water.

The Passive Solar House, James Kachadorian (Chelsea Green Publishing, 2006)
Using solar design to keep your house warm and cool.

The Renewable Energy Handbook, William H. Kemp (Gazelle Drake Publishing, 2006)
A US take to generating your own energy in a rural setting.

Tapping the Sun, Chris Laughton (Centre for Alternative Technology Publications, 2006)
A practical guide to heating your water from solar power.

The Woodburner's Companion, Dirk Thomas (Alan C. Hood, 2006)
A US guide on practical ways of heating with wood.

Browse

Solar Cooking Archive (www.solarcooking.org)
Loads of information, including 'how-tos' on cooking with solar power.
Solar Online Australia (www.solaronline.com.au)
Typical wind- and solar-power suppliers in Australia.

Solar Power Answers (www.solar-power-answers.co.uk)
Rather home-grown-looking site, but a lot of information as well as sales.

Wind Trap (www.windtrap.co.uk)
Source for small wind turbines, inverters, wind-up torches etc.

Wind-Works (www.wind-works.org)
Canadian site with wide range of information and books on wind power.

Wood-burning stoves
You will find a wide range of alternative-fuel stoves by putting wood stoves into a search engine like Google.

The Staff of Life

Finding and making the most of your food and water.

Read

Bush Tucker Field Guide, Les Hiddins (Explore Australia, 2002)
Living from the wild, Australian-style.

The Calorie, Carb and Fat Bible, Juliette Kellow and Rebecca Walton (Weight Loss Resources, 2007)
Comprehensive guide, one of many calorie-counter books available.

The Composting Toilet System Book, Carol Steinfeld and David Del Porto (Ecowaters, 2007)
A practical guide to choosing, planning and maintaining composting toilet systems.

The Concise Guide to Self-Sufficiency, John Seymour (Dorling Kindersley, 2007)
Simplified and compact introduction to living off your land.

The Drinking Water Book, Colin Ingram (Celestial Arts, 2006)
Aimed at making ordinary drinking water safe, but includes good information on alternative ways to purify water.

A Field Guide to Edible Wild Plants, Lee Allen Peterson (Houghton Mifflin, 2000)
A good guide for eastern and central North America.

The Forager Handbook, Miles Irving (Ebury Press, 2007)
An excellent guide to finding and living off edible plants in Europe.

How to Store Your Garden Produce, Piers Warren (Green Books, 2003)
  There's not much hope of living off the land if you can't keep your produce through the winter: this guide shows you how.
Outdoor Survival Handbook, Ray Mears (Ebury, 2001)
  Although focused on outdoor living, includes plenty of information on living off the land.
New Complete Self Sufficiency, John Seymour, Will Sutherland and E. F. Schumacher (Dorling Kindersley, 2003)
  Detailed guide to living a self-sufficient life with typical DK illustrations.
Nutrients A–Z, Michael Sharon (Carlton, 2005)
  User's guide to vitamins, minerals, supplements and more.
Sprouts and Sprouting, Valerie Cupillard (Grub Street Publishing, 2007)
  Instructions on sprouting and seventy recipes for using sprouts.

Browse

Composting Toilet World (www.compostingtoilet.org)
  The name says it all.
Cost Plus Water (www.costpluswater.com)
  Informative US site selling water-purification products.
Excel Water Technologies Inc. (www.excelwater.com)
  US site with a good range of water-purification technologies.
Nutrition Essentials (www.eatwell.gov.uk/healthydiet/nutrition essentials)
  UK government site detailing healthy eating and vitamin/mineral requirements.
Precision Water (www.precisionwater.com)
  Canadian water-purification vendor, selling distillers as well as filters.
Travel With Care (www.travelwithcare.com)
  UK travel site with a good range of water-purification products (also anti-mosquito etc.).

Wild Weather

The essentials for coping with drought, cold, flood and storm.

Read

Drought Resistant Gardening, Royal Horticultural Society (Dorling Kindersley, 2003)
Keeping your garden alive with drought-resistant plants.

Hurricane Almanac, Bryan Norcross (St Martin's Griffin, 2006)
Labelled 'the essential guide to storms past, present and future', includes guidance on preparation and survival.

Lightning Strikes: Staying Safe Under Stormy Skies, Jeff Renner (Mountaineers Books, 2002)
While primarily aimed at climbers, gives good advice on lightning survival.

Living with Wildfires, Janet C. Arrowood (Bradford Publishing Company, 2003)
Prevention of, preparation for and recovery from wildfires.

The UV Advantage, Michael F. Holick (iBooks, 2005)
How to make good use of sunlight to attain correct vitamin D levels.

Browse

Air-conditioning
You will find a wide range of air-conditioning products by putting air conditioning *or* air conditioner *into a search engine like Google.*

Hurricanes (hurricanes.noaa.gov)
Information from the US National Oceanic and Atmospheric Administration.

Lightning Safety (www.lightningsafety.noaa.gov)
Guidance from the US National Weather Service.

Living with Drought (www.bom.gov.au/climate/drought)
The Australian Bureau of Meteorology's guide to drought.

National Drought Mitigation Center (www.drought.unl.edu)
Good information on drought from this US site.

Safe and Well

Find out more on personal security and health.

Read

First Aid Manual, St Andrew's Ambulance Association, St Johns Ambulance, British Red Cross (Dorling Kindersley, 2006)
Detailed first-aid guidance from the professionals.

First Aid for Babies and Children Fast, British Red Cross (Dorling Kindersley, 2006)
Specialist advice on dealing with health problems and injuries of children.

Home Security, Heather Alston and Calvin Beckford (New Holland Publishers, 2005)
Common-sense guide to securing your house and garden.

The SAS Self-Defense Handbook, John 'Lofty' Wiseman (Lyons Press, 2000)
A well-structured introduction to self-defence from an expert.

The SAS Urban Survival Handbook, John 'Lofty' Wiseman (Harper Collins, 1996)
A general guide to surviving everything that city life can throw at you.

Secrets of Street Survival: Israeli Style, Eugene Sockut (Paladin Press, 1995)
Only consider reading this book if you are prepared to carry at least two handguns.

Browse

Canadian Red Cross (www.redcross.ca)
Information and first-aid courses across Canada.

First Aid Action (www.bbc.co.uk/health/first_aid_action)
Interactive online first-aid course from the BBC (also details courses).

A Guide to Home Security (www.crimereduction.gov.uk/cpghs.htm)
UK government crime-reduction site.

Home Security
You will find a wide range of home-security products by putting home security into a search engine like Google.

iFirstAid (www.sja.org.uk/firstaid)
Audio first-aid guidance from St John Ambulance (also details courses).

Life Axe (www.lifeaxe.com)
A device for breaking through double-glazed windows to get out in an emergency.
Red Cross Australia (www.redcross.org.au)
First-aid courses across Australia.
St John Ambulance Australia (www.stjohn.org.au)
Information and first-aid courses across Australia.
St John Ambulance Canada (www.sja.ca)
Information and first-aid courses across Canada.
Self-Defence Strikes (www.cse.dmu.ac.uk/~bb/dg/SelfDefence)
Useful home-grown self-defence guide.
Urban Survival Tools (www.urbansurvivaltools.com)
US urban survival kit retailer.

Keeping Our Humanity

Further reading and links on stress-relief, problem-solving and making the most of your time.

Read

Anyone Can Tell a Story, Bob Hartman (Lion Hudson, 2002)
Advice on the art of storytelling.
Instant Creativity, Brian Clegg and Paul Birch (Kogan Page, 2006)
A sourcebook containing a wide range of techniques for generating ideas. Also available as a downloadable ebook from www.cul.co.uk/titles/instantdownloads.htm
Imagination Engineering, Brian Clegg and Paul Birch (Pearson Education, 2000)
A developmental course to help build your personal creativity and ability to solve problems.
Instant Time Management, Brian Clegg (Kogan Page, 2001)
Quick techniques for time management – ideal if you don't have time to manage your time.
Story, Robert McKee (Methuen, 1999)
Although specifically about screenwriting, a powerful analysis of the nature of story.

Thinkertoys, Michael Michalko (Ten Speed Press, 2006)
 Inspiring techniques to improve your personal creativity.
Total Stress Relief, Vera Peiffer (Piatkus, 2005)
 Practical techniques to protect yourself from stress.

Browse

Business Balls Time Management (www.businessballs.com/time.htm)
 Useful tips on keeping track of your time.
Creativity Unleashed (www.cul.co.uk)
 Information and help on being more creative. Click on the Information Site tab.
DeStress (www.cul.co.uk/destress)
 Tips and advice on stress-control.
GoCreate (www.gocreate.com)
 Creativity advice, techniques and information.
Mind Tools Stress Management (www.mindtools.com/smpage.html)
 A good collection of stress-relief tips.
Stress Busting (www.stressbusting.co.uk)
 Lots of information on this site from the New Age end of the stress business.